ACPL ITEM
DISCARDED

YO-AAW-446

.994 H36i
edus, Tibor J.
indicted: cancer research

INDICTED:
CANCER RESEARCH

FRAUD, FEAR, FUTILITY, AND THE $150,000 MOUSE EXPOSED AT LAST

Tibor J. Hegedus, Ph.D.

Avondale Laboratories, Inc.
Avondale, PA

Allen County Public Library
900 Webster Street
PO Box 2270
Fort Wayne, IN 46801-2270

Indicted: Cancer Research. Copyright © 1993 by Tibor J. Hegedus, Ph.D. Printed and bound in the United States of America. All rights reserved. No part of this book may be reproduced in any form by any electronic or mechanical means including information storage and retrieval systems without written permission from the publisher, except by a reviewer, who may quote brief passages in a review. Published by Avondale Laboratories, Inc., 252 Hillendale Road, Avondale, PA 19311.

Publisher's Cataloging in Publication

(Prepared by Quality Books Inc.)

Hegedus, Tibor J., 1952-
    Indicted--Cancer Research: fraud, fear, futility and the $150,000 mouse exposed at last / Tibor J. Hegedus.
    p.   cm.
    Includes bibliographical references and index.
    Preassigned LCCN: 92-97251.
    ISBN 0-9634721-9-4

    1. Cancer--Research--United States--Popular works.
    2. Cancer--Research--Controversial literature.   I. Title.

RC267.H44 1993   616.994'007
                                      QBI92-20158

ACKNOWLEDGEMENTS

Thanks to all who have asked meaningful questions about cancer and cancer research. Your questions and comments showed me how alarmed and intimidated you are by this disease, yet determined to understand it. You guided my motivation to set the record straight.

Peter Ciullo took a personal interest in this project and made significant editorial contributions that molded the final product. In addition, his patience and welcomed advice is truly appreciated.

A number of friends and neighbors kindly took the time and made the effort to read and to critique my book. Their observations also contributed to producing the final product.

This book is the result of a community effort called the typical American.

TABLE OF CONTENTS

exposure rate. It sounds shocking, but it's true. The process just described is pregnancy. Sperm is the mutagen, and an ovum is the target cell. No technical degree or great leaps of faith are required to understand this process. Likewise, you can understand the process of cancer formation, or carcinogenesis.

PREFACE

Confusion has its costs. When these costs concern cancer, they easily can reach the **maximum** that includes the total of your life savings and even your life. You hired cancer researchers because they claimed to have the necessary experience and expertise to **minimize** your costs by preventing and curing this disease. What happened? Have they satisfactorily answered any of your questions, or alleviated your fears?

Responses, disguised as answers, are shrouded in shades of gray that intentionally contribute to your confusion, multiplying your cares. After the investments you made into cancer research, you deserve definitive answers about cancer related to you in an easy to understand style. With understanding comes the control to make proper decisions. With control comes the elimination of fear. The technical language of scientifices should not cause you to lose control by being incompre hensible. While typical discussions about cancer ar designed to be presented in highly technical jargon, yo can understand this disease. All you really need is prope information and common sense.

For example, there is a potent mutagen that targets single cell in humans. If it enters this cell, a cascade biochemical events result immediately that stimulate cell division which can lead to a growth of about 7 pound within 38 weeks. Up to 75 million Americans are expose to this mutagen every year. However, with the odds only 1 in 500 million that it will enter the cell, repeat exposure can occur for years before any effects are fe On the other hand, the first exposure can initiate th growth. This process occurs to about 4 million America each year according to the U.S. Census Bureau. government official will make an attempt to reduce t

INTRODUCTION

Ever since I joined the ranks of cancer researchers, people have asked me why there is no cure for cancer. Many have offered their own theories. Some believe in a government conspiracy to withhold the cure in an attempt to control the human population. This reasoning is based on the belief that overpopulation will result if a steady, predictable number do not die of this disease each year. Others know that treating cancer is a multi-billion dollar a year industry bringing physicians, hospitals and pharmaceutical companies huge profits. Any cure for cancer would reduce these profits, and run contrary to the best financial interests of these groups. Still others believe that curing cancer is just too complex a problem with no resolution, so we should not waste our time and money on this pursuit any longer.

Likewise, many have shared their theories on the mechanism for the cause of cancer. Much of what they have expressed is based on information supplied by the news media. Of course, their information is based upon what "noted" researchers in the field have told them. Some believe that viruses cause cancer. Others believe cancer is due to heredity. Many are convinced that chemicals, radioactive compounds and ultraviolet radiation cause cancer.

Regardless of prevailing theories and popular beliefs, however, everyone wants to know with certainty what actually causes cancer and if cancer can be cured.

About the Book

One reason for writing this book is to answer these questions. You have the right to know how your tax dollars and charitable contributions to cancer research organizations are being spent. This book is a rare opportunity to examine all aspects of cancer research from an insider's view, written in a style the average person can easily understand.

Everyone is concerned about what is referred to as the trilogy of cancer: 1. Prevention, 2. Cause, and 3. Cure. Cancer has been recorded as a human disease since the ancient Egyptian physicians first described this condition at about 3000 B.C. It is a logical assumption, therefore, that after 5,000 years at least one part of the trilogy should have been solved. Yet with the growing number of cases each year, it is painfully evident that cancer researchers have failed in their attempts to explain any part of the trilogy. One fact is certain, no person is immune from developing cancer.

In this book, no organization or type of cancer researcher will be spared criticism of their work to date. This work includes data presented and all theories on the mechanism of cancer. You are correct if you believe that there is a conspiracy in cancer research. However, this conspiracy is not rooted in official federal government policy on human population management, or even in corporate greed. This conspiracy is more insidious than any of you have imagined. The very institution of cancer research is as malignant as the disease it is studying.

This conspiracy does not arise primarily from economic reasons, although many cancer researchers do enjoy an enviable lifestyle. They work in a very clean and pleasant environment. Most of the heads of cancer research programs throughout the world make over $100,000 per year and they have virtually unlimited vacation time which is disguised as attendance at seminars or meetings in the field. Perhaps the salary is not high when compared to athletes and entertainers, but how many of you earn a six-figure income? These individuals live a very comfortable life and enjoy many perks as well as job security. Many of you are not so fortunate. Yet you are coerced into financially supporting them. This coercion is slick and subtle taking the forms of your tax dollars, charitable contributions and purchases of pharmaceutical products. You are not getting your money's worth.

It is not money but ego that is the major motivation of these researchers. Each one of them wants to take credit for discovering either the mechanism of cancer or the cure in order to win the ultimate reward - a Nobel Prize. In obtaining this award, they believe their name will live forever in an "Einsteinian" manner. In fact, two molecular biologists already have won a Nobel Prize in medicine for research that led to the development of the oncogene theory of cancer. The awarding of this prize was disputed by a third molecular biologists who claimed to have originated the theory. This controversy was reported in the major newspapers. As you will read later, this theory cannot be substantiated because it has no basis in fact - but what encouragement to other Nobel Prize aspirants! This example serves as a documented case of how cut-throat cancer research is conducted daily at your expense.

To obtain the Nobel Prize, most researchers will stop at nothing. The destruction of a young researcher's career after he has provided his supervisor with valuable information is the despicable norm. One researcher will steal ideas from another researcher, especially a subordinate, and claim it as his own. There is no legal recourse for the subordinate. Researchers have suppressed data that contradicted work they previously published. The ego of many of these researchers is insatiable. You should know *that it is more important for you to suffer and die than it is for them to get their egos bruised.* It is important for you to understand that the thieving supervisor is intellectually incapable of pursuing the ideas stolen from the subordinate. If the grand guru were truly competent, then he would not have to resort to the theft of ideas from anyone. The bottom line is that you pay for this continuous criminal behavior with your money, and with your lives. This carnage must stop.

About the Author

Let me introduce myself. My name is Dr. Tibor J. Hegedus. After receiving my doctorate in Medicinal Chemistry, I spent more than three years working with several internationally recognized cancer researchers. As a result of this experience, I had access to much of the "behind the scenes" activity and dialogue denied to the press and public. I already had been involved in cancer research for about 10 years prior to working with these people. It is my commitment to work in cancer that was always part of the reason I was hired.

When I was an undergraduate pre-med major, I was an "armchair" cancer researcher like many of you. I read the newspapers and listened to the news for information on this subject. At that time it seemed that a major breakthrough in this area would occur very soon. Most of the data and theories made sense to me at that time. To improve my chances of getting into medical school, I worked for several months as an orderly in Lankenau Hospital back in 1977. One of my responsibilites was the care of terminally ill cancer patients. I helplessly had to watch them suffer and then ultimately die because there was no cure. Some of the patients were so ill that not even the most potent pain relievers had any effect. I began to believe then that if I ever got a chance to become involved in cancer research, I would make a significant contribution. Watching others suffer and die of cancer was difficult enough for me, but fate had a more personal tragedy in store when my mother became diagnosed with inoperable cancer.

A brief medical history of my mother's last two years describes my first exposure to the incompetency of physicians in general and cancer researchers in particular. One Sunday in May of 1977, my mother suffered cardiac arrest just as she was to enter work. Fortunately, she was rushed quickly to Lankenau Hospital within about five minutes. Her condition was stabilized by the time I arrived. She was alert but clearly frightened. I assured her that she would be alright. The attending physician explained her condition to me and the necessity to keep her overnight for routine observation. Within 24 hours of admission, she suffered a stroke and became paralyzed completely on her right side. Due to the careful attention and skill of our personal physician Dr. Bodi, she was able to walk and talk again in about two months. However, she kept complaining about a severe pain on her right side. The pain was from gallstones, so she had an operation to remove her gallbladder less than a year after her stroke. About three months after her gallbladder operation, she became jaundiced. Tests revealed that she had "inoperable" cancer of the common bile duct. This duct leads directly from the liver to the gallbladder.

How could the surgeons performing the gallbladder operation not have seen the tumor? The physicians at Bryn Mawr Hospital who diagnosed her cancer were not the same physicians who had performed the gallbladder surgery at the now defunct hospital called West Park. The physicians at Bryn Mawr explained to me that my mother had only about a year to live because this type of cancer grew rapidly. In my opinion, there had to have been a tissue mass large enough for the surgeons to detect and to remove at the time of the gallbladder operation, saving my mother's life.

Radiation therapy was used to treat the tumor. Surgeons at Bryn Mawr Hospital implanted metal plates around the tumor to make focusing of the radiation beam easier and more accurate. The surgeons who were to perform the operation told us that this procedure would also minimize the development of side effects such as nausea and vomiting that are common with radiation therapy. Even with this radiation focusing technique, my mother still suffered a number of side effects including nausea, vomiting and weakness. During the last year of her life, I watched my mother go from a weight of about 150 lbs. to about 60 lbs. Often she was in such pain that potent analgesics (pain killers) could not alleviate her suffering. My mother died almost one year from the day diagnosed, as predicted by the physician at Bryn Mawr Hospital.

A few days after my mother's death, I entered graduate school determined to learn as much as I could about cancer. I studied theories on the mechanism and possible cures. I took the courses I considered necessary to obtain the background required for the successful development of cures for cancer. I studied pharmacology, toxicology, biochemistry, teratology, analytical toxicology, medicinal chemistry, computer programming and spectroscopic identification of organic compounds. For my graduate thesis, I designed and synthesized compounds that laid the foundation for future development of anti-cancer drugs. Of course, further development of these compounds was required before they could be tested in humans. But the important feature about my compounds was that I already was working toward anti-cancer drugs based upon a novel strategy compared to existing approaches. In a later chapter, I review for you most of the different methods of cancer treatment currently being used and why each necessarily fail based upon proven scientific principles. Current strategies are based upon a combination of wishful thinking and outdated theories.

When I completed my graduate studies, I felt confident about my abilities to design and to develop any type of drug. My main interest, however, was in the design of anti-cancer drugs. While I was a graduate student, I noticed that part of the problem in cancer research was that inadequate model systems were being used to examine this disease. Either whole animals, such as mice and rats, or microorganisms, such as bacteria, were used in attempts to learn more about the mechanism of cancer or to evaluate new drugs. Each of these model systems have important problems. Whole animals are very complex from a biochemical point of view. When a chemical is administered to a live animal, a number of reactions occur which makes it difficult to precisely monitor how that chemical is processed. Similarly, the conclusions made based on the study of bacteria often can not be generalized to the complex whole animal.

Fortunately some researchers were working on the development of an intermediate type of model system that could help to resolve these problems. One system is the tissue culture of cancer cell lines. Tissue culture means that cells comprising a given type of organ can be grown in a plastic flask. The cells are placed into a nutrient rich "broth" much like beef boullion. After graduate school, I had the opportunity to work with an internationally recognized cancer researcher who had human lung cancer cell lines growing in culture. I immediately accepted the offer to join this researcher because I knew this was the ideal test system for learning more about the biochemistry of cancer cells.

The title of my position was post-doctoral research associate funded by the Center of Excellence. I worked at the Oak Ridge National Laboratories in Oak Ridge, Tennessee and at the University of Tennessee College of Veterinary Medicine, Department of Pathobiology in Knoxville, Tennessee. Not only did I have the opportunity to work with cancer cell lines, I also had the opportunity to apply what I observed in tissue culture to a whole animal, which was the Syrian golden hamster. The tissue culture work was a way to simplify the biochemistry of the whole animal. I was able to go back and forth between cancer cells in culture and the whole animal in the application of a new theory on the mechanism of cancer.

To continue my work, I next accepted a position with Bionetics Research, Inc. I was not allowed to work with tissue culture there. I found ways to continue my work by using isolated enzymes and enzyme systems to understand better the biochemistry of cancer cells versus their normal cell counterparts. While working for this company, I had many occasions to work with and meet some of the top cancer researchers in the world, including those from the National Institutes of Health (NIH), the National Cancer Institute (NCI) and the Food and Drug Administration (FDA). I had the opportunity to observe and to evaluate their technical expertise and their personal qualifications as well as the political and ethical considerations in the working environment. All of these factors drastically shaped my perception of cancer researchers. As you will read, science and cancer victims are the hapless losers.

Here we are more than 15 years since my mother's death and no real new advances exist. Researchers still cannot solve any part of the trilogy of cancer, despite grandiose claims to the contrary. I am sick and tired of hearing yet one more story on the news about "breakthroughs" and "promising new leads" that are all hype and no hope. I'm fed up with bizarre new "treatments" and wacky new theories designed to get a scientist more often in the papers, and more of your money into his pockets. In fact, I've devoted an entire chapter of this book to explaining how you can conduct cancer research right in your home that is just as valid and just as valuable as the junk that is being turned out by the "experts" today. This is not meant to be cute, but rather to impress you with just how easily and arrogantly the cancer research industry trades false hope for hard cash. In short, I want you to be as angry and disgusted as I am.

About Hope

What I don't want is for your disgust to foster dismay. I do not want you to exchange false hopes for no hope. There is hope because the talent is available to turn that hope into reality, cures for cancers - if only allowed. This book is an inside view of the diseased spectre of cancer research. As such it is, and should be, disquieting if not actually revolting. But more than an expose, this book prompts you to consider what will be done, once the right persons are given the job.

CHAPTER 1

THE $150,000 MOUSE

Those of you not involved in research using animals probably are not aware of the costs and procedures associated with buying an animal. Let's assume that you have decided to start your own research center to discover an anti-cancer drug. Because any drug you develop must be tested in animals before it can be prescribed for humans, you decide that it would be best to begin your work by purchasing a single mouse. This seems to be a logical decision in that you want to keep initial costs at a minimum in case your efforts do not prove to be successful. As I reveal the actual price of a mouse for research, please keep in mind that this process is going on every day, and that you are paying for it. Remember, the research in a hospital or university is supported by federal funds you pay through tax dollars. If the research is by a private pharmaceutical company, you pay for it in increased prices for their products.

3 1833 02478 7738

Of Mice and Money

Before you learn how to buy a mouse, you need to understand what a federal grant is. A scientist puts together a proposal asking the government for money to conduct research in a given field of interest. This proposal is reviewed by a committee of experts in the field. If the committee likes the proposal, money is given to the scientist. The money so awarded never has to be paid back. This money is not a loan. Salaries are part of the justifiable costs of the research project. For example, a physician who is earning income in private practice can also give himself a salary from the grant money separate from and in addition to the salary collected from his practice. A college professor receives a salary from the school and can collect another salary from the grant. The scientist can not directly use grant money to buy a new BMW or take a one month vacation to tour the world. But that extra $30,000 to $50,000 salary can help to purchase these "necessities" of life.

Let's say that you have obtained money to buy a mouse for research. This money can come from either a grant, or contributions from individuals, much like the money you may give to one of a number of cancer related organizations. Before you buy your mouse, you must first decide which of several strains of mice you need. A strain is similar to a breed of dog. Each strain of mouse has been genetically engineered to possess different characterisitics. For example, there is a "nude" mouse strain which has a drastically reduced immune system. This strain is very prone to catching infectious diseases. When you write your research grant proposal, you must justify the reasons for choosing a particular strain. To help you decide which mouse to use, I suggest that you go to a medical school library and read several recent articles in cancer research magazines to learn which strains are currently fashionable among the other cancer researchers. By reading these articles you also will learn how to phrase your proposal in the way most appealing to a review committee.

Once you have selected a strain of mouse, it must be purchased from a federally licensed dealer. Names and addresses of dealers will be listed in the articles you have read. A mouse will cost about $5. This would not appear to be much money. You might think that all you have to do now is buy a cage, some food and a water bottle. It would seem that you are set to start your own cancer research at a cost of about twenty dollars. I am sorry to inform you that your proposal will not fly. You have not considered all of the requirements for housing a mouse.

In order to house a mouse, you must follow rules established by the Food and Drug Administration (FDA). This procedure requires that you have a full-time veterinarian on staff to ensure that the mouse will be humanely treated at all times. In essence, this means that the veterinarian will approve of the way you eventually kill the mouse and dispose of its body. You are not allowed to throw out the dead mouse in the weekly garbage either. The veterinarian's salary could run between $30,000 to $100,000 per year. We can average this salary to $50,000. You have to keep in mind that while the veterinarian is looking after your mouse, he or she can not be in private practice. You need to compensate this professional for lost income.

The next step is to purchase a cage. You can not go to your nearest pet shop and just buy one. Those cages are not federally approved. You must purchase all of your supplies from companies specializing in scientific products. You can buy a mouse starter kit for about $81.00. The supplier of this kit also is listed in the articles on cancer research. This kit includes a plastic cage about the size of shoe box, a metal lid for the cage, a glass water bottle, a rubber stopper for the water bottle and a bent stainless steel tube to put into the water bottle so that the mouse can drink. The metal lid serves to keep the mouse in the cage and as a container for food.

The mouse needs bedding for the cage. Again, you can not go to the local pet shop for bedding. The contents of the bedding must be certified free of chemicals that might affect the results of your experiments. Pesticides, for example, may be present in the bedding bought from the pet shop. However, the bedding purchased from a scientific supply company can not be used straight from the package. It must be sterilized first. Microorganisms may be present in the bedding; these can affect the outcome of your experiments.

Likewise, you can not use water from the tap. The water you give to the mouse must be double distilled and triple filtered. There are far too many chemicals present in tap water that can affect the results of your experiments. Ordinary tap water contains chlorine, iron, copper, cobalt and possibly minute amounts of bacteria. Using the same rationale, you must purchase laboratory mouse chow, which also must be sterilized.

Water can be purified using any one of several available purifying systems. The prices for these systems range from $450-$7,000 depending on the various features of each model. It is like buying a car. Prices vary due to make and model. You could get by with a mid-range water purifying system for about $2,000. To sterilize the water, feed, and supplies, you have to buy an autoclave. Autoclaves also come in different models depending on features. One mandatory feature is the ability to sterilize fluids, such as water, as well as equipment. You can expect to pay $31,000-$37,000 for an autoclave. Again, we can pick a mid-range model for about $35,000.

Laboratory animals function according to circadian rhythms. That is to say, mice are more active at certain times of the day or night than at others. This pattern is also true for bodily functions such as the digestive tract. Due to these circadian rhythms, mice have to be housed in carefully controlled conditions, which include temperature, humidity and cycles of 12 hours of light followed by 12 hours of dark. It will cost you about $25,000 to have a room suitably renovated to house your mouse.

If you set this room by itself somewhere in the middle of the country, then you will not have be concerned about unions. However, your room most likely will be located in a research facility, such as a school or government laboratory. In this case, you have to hire union personnel to maintain and operate the major pieces of equipment and to feed your mouse. A technician's salary is about $30,000. You should expect to pay at least $1,000 per month for electricity because the heating/air conditioning system, the humidfying system and light timer control must operate 24 hours a day every day of the year. Typical experiments on mice last for a year, so you will pay about $12,000 in electrical bills. Let us assume that you need to spend $1,200 for the year on miscellaneous items such as laboratory mouse chow, notebooks, telephone bill, etc. The total price tag for your first five dollar mouse comes to $155,086 for the year, see Table 1-1.

Of course the more mice you buy, the lower the cost per mouse. You will not have to purchase major pieces of equipment again. And for the same salary, the veterinarian you hired can see to the welfare of several hundred mice instead of one. If you purchase 1000 mice at a price of $5 each your expenses per mouse can be reduced to about $170/mouse per year.

In the calculation of cost per mouse, I have assumed that you already own a building in which to house the mouse. If this is not the case, then add to the cost of a

Table 1-1. Summary of Costs Associated With Buying
A Mouse for Cancer Research

Item	Price
Mouse	$5
Veterinarian	$50,000
Cage kit	$81
Water purifying system	$2,000
Autoclave	$35,000
Renovations	$25,000
Technician's salary	$30,000
Utilities	$12,000
Miscellaneous, including food and bedding	$1200
Total	**$155,286**

mouse the price of the building. Once you buy your mouse, you will have spent over $155,000 dollars. You still have not prepared a chemical which could be used as an anti-cancer drug. But you do own one happy, pampered mouse living better than royalty.

At one cancer research facility where I worked, the group leader had over 8,000 mice and over 2,000 rats killed in one year. You paid $40,000 for mice and $16,000 for rats that year. Rats cost about $8 apiece. That year you spent $56,000 on mice and rats for just one researcher. I can not say whether you are satisfied with your purchase, but I am not thrilled. This researcher worked at that facility for at least 15 years. By factoring in inflation and staying on the conservative side, the average yearly cost for mice and rats was $35,000. In 15 years, you bought $525,000 worth of mice and rats for this researcher. All of these carcasses have been cremated. This means that you spent over half a million dollars for about 50 pounds of ashes without knowing it. Of course this cost does not include equipment purchased and the other items I discussed. I am sure you would agree that the money would have been well spent if that researcher could tell us what the mechanism of cancer is, or present us with a cure for any cancer. However, the reality is that this individual can no more explain the mechanism of cancer today than he could 15 years ago. All this individual can do is list some chemicals and tell you which cause lung cancer, which cause lung damage with no cancer or which cause no damage to the lung. I do not feel this researcher gave me my money's worth.

In addition to mice and rats, cancer researchers have dosed frogs, guppies, trout, dogs, goats, salamanders, chickens, coyotes, and many other species of animals with chemicals. These chemicals have been labelled either as chemcial carcinogens or anti-cancer drugs, depending on the preference of the researcher doing the work. Millions of your dollars have been spent to develop what are termed "genetically pure" strains of mice with the thinking that these strains will lead to a better understanding of the mechanism of cancer. If they need genetically pure strains of mice, then what is the logic of using guppies and coyotes, which are not so genetically pure?

Remember, this is only one researcher at one facility spending this money without obtaining meaningful results. There are several thousand researchers and several hundred research facilities. It's very easy to understand how billions of dollars are spent each year on cancer research. Why the results are lacking will be discussed in subsequent chapters of this book. One aspect of cancer research which is very irritating is that almost all of these researchers compete with each other, even though they are paid to cooperate with each other. We are forced to support these petty, childish minds to compete with each other in the laboratory while millions of people suffer and die of cancer each day worldwide.

Sacrifice for Research

Let me help you to understand even better how animals are used in cancer research. Two types of studies are conducted, either short-term or long-term. In a short-term study, researchers look at the response a given carcinogen elicits inside of a whole animal within the first few hours of administration. In a long-term study, they want to know in which organs tumors form, or in the case of a potential anti-cancer drug, which animals survive the longest. At the end of both types of study the animals always are killed.

To better understand how a carcinogen is acted upon by the body, the liver often is removed after the animal is sacrificed. The scientific word for butchering animals is sacrifice. The liver is used because it the major organ in which metabolism occurs. This metabolism has a different meaning than what you are probably familiar with in the context of dieting. Metabolism to a biochemist is the way that enzymes alter chemical structures. For example, if you drink a beverage containing alcohol, the liver metabolizes the ethanol in the drink. The next day your urine will contain acetic acid, which is vinegar, as a result of this biochemical activity in your liver. In a similar fashion, chemical carcinogens are metabolized by the liver and other organs. Researchers want to understand the chemical transformations of carcinogens in order to understand the mechanism of cancer. For this type of experiment a researcher uses one rat or two mice per day.

Long-term studies are divided into two parts. In the first, researchers determine how much of the carcinogen to give the animal, because it will be dosed about every two or three days for several months. They do not want to give the animal too much of the carcinogen. At too high of a dose, the animal will die from the chemical before it has the opportunity to form tumors. In order to determine how much chemical to give to the animal, about 50 animals are divided into several groups. Each group receives a different amount of the chemical in one "shot". The range is from a low dose up to a very high dose. The animals are then simply observed until they either die or survive. A head count is made of the deaths. The dose at which 50% of the animals die is used as the basis for long-term dosing of the other animals. This method for determining the lethal dose at which 50% of the animals die is called the LD_{50}. The experimental dose given to the animals during the long-term study is typically set at 10% of the dose that killed half of the animals. Long-term studies usually last for one year. As often as cancer researchers can get away with it, they use the highest dose of a chemical that will allow at least some animals to survive long enough to develop tumors. The animals used in the first part of the study are killed and not used for the second part of the long-term study.

Money for Nothing

Now you have an understanding of how and why cancer research can be very expensive. Housing any laboratory animal is expensive. Some laboratory animals, such as chimpanzees, can cost $10,000 apiece. You have an idea of how animals are used in cancer experiments. Despite all of the work generated by these studies and the billions of dollars spent, the researchers still can not tell you how cancer forms or provide any cure. It's time for you to learn why this situation exists by taking a closer look at who conducts these experiments.

CHAPTER 2

THE "PROFESSIONAL" CANCER RESEARCHER

So far I have discussed cancer researchers in a rather generalized way. These individuals come from three different backgrounds. They are primarily physicians, biologists and chemists. Biologists in cancer research are most often molecular biologists who examine the location of genes. Chemists include biochemists, who look at how and where carcinogens are metabolized, and synthetic organic chemists, who develop compounds to be tested as anti-cancer drugs. People with other backgrounds may provide supportive roles, such as analytical chemists who define and analyze for chemicals; and pathologists who locate the formation of tumors.

We will concentrate on physicians, molecular biologists and chemists because they are the ones conducting most of the research in cancer. I will show you not only why they have failed to cure cancer, but, in addition, why these individuals do not have the appropriate credentials to cure this disease, unless they get lucky.

Professional Con Artists

I think that I can best sum up the history and current practices in cancer research by using an analogy. We all agree that a major environmental concern for us is the burning of fossil fuels like gasoline to move our vehicles. We can also agree that it would be in the best interest of our planet if an alternate means existed of moving these vehicles without the use of fossil fuels. The heart of the problem is that all of our means of transportation require the use of propulsion. A jet moves because fuel is burned in the engine that turns a turbine at high speed thrusting gases out the back of the engine and against the air, thereby moving the jet forward. A car moves forward by the same principle of propulsion using the friction between the tire and the road. Gasoline is the fuel that makes the engine work.

I could propose that we look for ways of moving vehicles which do not rely on the principle of propulsion. If successful, we would no longer have to use the fossil fuels that pollute our environment. Because I have identified the problem and have offered a solution, it is logical that I am the most appropriate candidate to work on the solution. All I want you to do is give me one trillion dollars to work on this project. For your money, I will publish my work in newly created journals that feature my work as evidence that I am in fact working on this problem. I will hire the best people I can find to assist me and present seminars on this subject in various cities throughout the world. However, I need at least 30 years to come up with the solution to the problem. And, I can not guarantee that I will have the solution even then, or ever.

Nobody would accept these terms. Can you imagine going to your boss with the certainty of not being able to perform the task you were hired to do, and keeping your job? As a parent with a sick child, you can not wait indefinitely to take action to alleviate your child's illness. The normal world does not function according to this absurd notion - give me as much money and time as you can, but expect no guarantee of a return for your investment. Yet this is exactly the situation you are forced to support financially in cancer research. Con artists posing as cancer researchers do make a comfortable living by offering you nothing but hope for your dollar.

Almost all cancer researchers are con artists. They use very technical, incomprehensible jargon to impress you with their "accomplishments". This language is an intimidating smokescreen designed to hide their consistent failures. Where is the delivery of the goods they promised? Certainly after more than 5 decades of government sponsored research we can expect a solution to at least any one part of the cancer trilogy. These con artists seduce you into thinking something meaningful is being done. Every few days you hear a news story about scientists at some major cancer research center developing a promising new treatment for cancer, such as electromagnetic separation of leukemia cells from normal white blood cells. Another favorite line is that doctors have found a link between high voltage wires and leukemia and/or brain tumors in children. I guess that adults are immune to the carcinogenic effects of electricity. The list of such rubbish is tragically long.

If you are like most people, you do not understand and are not fluent in the language of "scientificese", if I may use that word. This is not to imply that you are unintelligient. It only means that there are limited areas of expertise for any given person. Most people are not interested in the medical sciences to the extent that I am or that other cancer researchers profess to be. Because you do not fully comprehend their language, you are easy prey for their tactics. You are being forced to buy "pigs in the poke". Unless someone like me, who has insight and personal experience in this field, is willing to share this knowledge, you will be forced to support these bogus research projects indefinitely. Even worse is the emotional trauma you may be forced to face someday by having your hopes built on false claims of a cure for your cancer.

Let me tell you now that *there is no cure for cancer.* Yet, cancer can be cured. This may sound like double talk right now, but I will show you later why this statement is true. First, you need to know why physicians, biologists and chemists can not unlock the mechanism of cancer or develop its cure.

CHAPTER 3

PHYSICIANS

Physicians have dominated the field of cancer research since they first diagnosed the disease in ancient Egypt. It would appear that they are the most logical group of individuals to examine this problem. Physicians are suppose to have knowledge about many human diseases and body processes. They also are suppose to be trained in the treatment of human diseases. Therefore, two logical questions to ask this group of cancer researchers is why they can't explain the mechanism of cancer and why can't they cure this disease.

The perennial battle cry from physicians is that they *need more money* to fund their research in order to solve these riddles. They periodically release to the press stories on current "advances" in either treatments of this disease or "discoveries" that shed new "light" on causative factors for cancer. I believe that physicians have had plenty of time and money to examine this problem. Let me describe for you how the United States government allocates grant money for medical research.

The United States is the only country in the world that I know of which considers the M.D. degree higher than a Ph.D. degree. Consequently, when there is a decision to be made between a physician or a Ph.D. receiving a research grant, almost invariably the physician gets the money. The main reason for this decision is that M.D.'s have a much stronger union, called the American Medical Association (AMA), than Ph.D.'s, who stupidly have no union to protect their rights. Physicians, due to membership in this corrupt union, hold many high positions on review committees that decide who gets grants. They look after their own first, and anyone else second. I know of several cancer research programs headed by physicians where the higher ranking Ph.D.'s must work for the lower ranked M.D. In these programs the real intellectual thrust of the program comes from the Ph.D.'s, but the M.D. takes the credit. Among the numerous examples to support my claim is a former head of the Biology Division at Oak Ridge National Laboratories who was merely a physician, with Ph.D.'s as subordinates. A physician likewise headed the Fels Institute for cancer and AIDS research at Temple University and had Ph.D.'s report to him. If you read the articles published carrying the names of these two physicians, credit is taken by them for initiating the work. However, the ideas and the experimental methodologies originated from the Ph.D.'s.

Degrees of Incompetence

Physicians are unqualified to head research programs because they earned only a nonresearching Masters degree. It is important for you to understand the difference between a researching degree versus a nonresearching degree. The order of ranking for college degrees is Bachelors followed by Masters and then the Doctorate or Ph.D. In the system for advanced degrees, one can earn either a researching or nonresearching degree. In order to complete the requirements for a researching degree, the student must decide upon an independent problem to be examined, conduct the experiments to find a solution to the problem, and report these findings in the form of a written text called a thesis. Laboratory work must be done in addition to completing a large number of classroom courses. To earn a researching degree, the student must demonstrate to the advisory committee a capacity to think and to be creative in problem solving. This criteria applies to both Masters (M.S.) and Doctorate (Ph.D.) degrees.

To obtain a nonresearching advanced degree, one merely needs to take a few extra courses instead of completing a research project. This criterion is typical of degrees in law, business administration and medicine. In taking these extra courses, the student simply has to demonstrate to the professor that a minimal amount of information was retained to pass tests. At no time is the student required to demonstrate his or her capacity to think. No problem is examined. This is the situation in medical school. The students take a carefully designed curriculum of courses throughout most of the four years of their education. In the last year and a half they make hospital rounds to put into practice what they learned in the classroom. These exercises are under close supervision by faculty members. The students examine a patient and make a diagnosis. They are either correct or incorrect. Diagnosing most human disease is not that difficult. There are only a fixed number of variables to be considered before the diagnosis can be made accurately. Often patients tell the physician what is wrong. There is, in short, nothing in the curriculum of medical school students that prepares them for the challenges of research. In fact, it is virtually impossible to flunk out of medical school, unlike every other academic program. There is no mystery as to why physicians have not been able to come up with either the mechanism of cancer nor any cure for the disease. They are inadequately educated to examine this problem.

For example, a medical school student is required to take only one semester each of biochemistry and pharmacology, two fields of study that are essential in unravelling the mechanism of cancer. I took two semesters of biochemistry and six semesters of pharmacology along with two years of pharmacology laboratory work. Even my education in these subjects is not equal to individuals who have Ph.D.'s in those areas of study. Biochemistry and pharmacology deal, respectively, with understanding life at the chemical level, and how drugs interact with biomolecules in the body. As you will learn later, cancer is both a chemical and a pharmacological event.

A fitting analogy that indicates the difference between the research capabilities of physicians and scientists is to compare the investigative abilities of a patrol officer to those of a detective. A patrol officer is certainly trained and educated in the procedures to follow at the scene of a crime. This person also knows something of the legal proceedings that follow during the investigation. However, the detective has additional training and education in the techniques of criminal investigation and is therefore better equipped to investigate the crime. A police officer may in fact have the ability to be trained as a detective, but that ability has to be demonstrated in a series of appropriate qualifying tests.In a similar manner, physicians learn scientific information and principles, but they have never been taught to use this information in a creative way to examine diseases. They certainly never had the experience to creatively solve novel problems in medically related research.

Physicians can fool you into thinking they are the best candidates for medical research because the lay person has neither the knowledge nor the resources to challenge them. Physicians speak of science and in the language of science, but an experienced mind can readily see how their theories are most often erroneous due to their limited understanding. Physicians can readily manipulate their data to arrive at predetermined conclusions because there is no one to challenge them except other cancer researchers, who will not do it. Other physicians and molecular biologists will not do so because they will not bite the hand that feeds them; chemists will not do so because they realize that rocking the boat spells career suicide. None of them anticipated my showing up. This lack of education is evidenced in the misguided and expensive, but worthless, research projects reported in their tabloids like <u>The Journal of the American Medical Association,</u> or (JAMA).

Typical Tactics
Romance

One tactic of physicians is to foster the romantic notion that they are tirelessly working long hours in a never ending quest for the cure of cancer. Those days of the lone physician working late at night for many years in his laboratory to find a cure for some disease are over. There is no physician today like Louis Pasteur, who discovered that infectious disease can be caused by microorgainsms contained in milk products. Nor is there another Fleming, who accidentally discovered penicillin while working late one night in his laboratory.

The cure for cancer can not be approached with this mentality because it is not an infectious disease. Yet the mindset of physicians is still back in the romantic past, hoping to make that great discovery in cancer which will astonish and delight the world. As a result, their approach to this disease is too primitive and outdated.

What you see coming from their research are such bizarre concepts as viruses that cause cancer, or that cancer cells are genetic mutations.

Likewise, since Pasteur showed the importance of the immune system in fighting diseases, physicians have simplisticly reasoned that they must merely increase the activity of the immune system to cure cancer. To do this, they argue, work must be done to learn more about the immune system. Another approach is to search for that wonderful "miracle" drug comparable to penicillin. This drug "obviously" must be contained in the immune system, so they offer you interleukins and interferons, which are isolated from leukemia cells. All of these strategies must, and do, necessarily fail.

I am not implying that physicians do not want cancer to be cured. They will still benefit financially because any anti-cancer drug will be available only by prescription. This means that you are forced to go to the physician and beg for the prescription of the drug that can cure your cancer at the going price. The physicians who are in cancer research want to be acknowledged as "the one" who found the cure. This focus on themselves is the driving force behind their involvement in cancer. There is a tremendous amount of competition between research-ers, regardless of educational background, to find either the mechanism or the cure. They battle each other behind the scenes. Officially, though, they present a unified front to the public and to the politicians who are responsible for funding their work.

Theft

Let me cite one of a number of personal experiences that demonstrate how the egos of physicians work. One physician in cancer research tried to suppress data obtained in a set of experiments conducted by his postdoctoral research fellow because this data contra-dicted findings the physician had published earlier. A Ph.D. colleague learned of these results and refused to let this work go unnoticed. This data led to a major discovery in understanding the mechanism of lung cancer induced by tobacco smoke. The Ph.D., due to status, forced this work to be published. The results of this experiment were consistently reproducible, which is rare in cancer research. As you may have guessed, the physician shared the credit for making this important discovery, which he initially tried to suppress, after the results of the experi-ment became generally accepted by other cancer researchers. Frankly, I have yet to read an article published by a physician that I consider to be valid scientific work.

Chemical Dosing

Another tactic physicians use is to mindlessly ram chemicals into animals at the highest tolerated dose and then report one of three results: 1. The animals developed tumors, 2. Tissue damage was observed with no tumors, or 3.There was no effect. This busy work is not science. There is no investigative strategy involved here. Just about any non-comatose human being could conduct this type of experiment as long as the procedure is described by a scientist in detail. It is important to determine which chemicals are carcinogenic. But, this work only provides a listing of carcinogenic compounds without giving further insight into the mechanism of cancer. It is much more important to know the mecha-nism that makes the chemical a carcinogen.

Misguided Sociological Studies

Another tactic physicians use to fool the public is to publish an article showing a link between cancer and something. It can be anything. As soon as I see or hear this, I know that the study is worthless. At best, these articles report poorly conducted sociological studies. There was a time when these studies impressed the public. Now, they only encourage a justified skepticism toward reports presented by all cancer researchers. Take for example a series of studies that linked cancer in women to sexual activity. One group of physicians reported that very sexually active women, such as prostitutes, have a higher chance of developing cancer of the cervix and uterus than less sexually active women. Another group of physicians published that nuns, who are sexually inactive, have a higher incidence of breast cancer than more sexually active women. If you believe the results of these two studies, then women are doomed. The only real choice women have is to decide to die from either cervical or breast cancer, unless they engage in the correct amount of sexual activity, whatever that may be. It is reports like these that widen the crediblity gap between cancer researchers and the public. And you pay for this rubbish.

There was a study reported several years ago that children who live around power lines have a 1.7 times greater chance of developing leukemia than children who do not live around power lines. More recently, some of these same researchers have tried to link living near power lines to brain tumors in children. This rate is not significant, if it even actually exists. The physician who reported this finding was awarded a $5 million grant to investigate this matter further and received his 15 minutes worth of fame in the process. Temporarily, his career was advanced. Power companies were forced to conduct their own studies. You paid for his career advancement twice. The first time with your tax dollars to fund his useless study and the second time with increased electrical rates to fund the power companies' own studies. There is no way this physician can tell you how electricity causes cancer. I found this study to be noteworthy in that electricity causes cancer only in children, not adults. I want this researcher to tell us how electricity can make the distinction between a child and an adult. What is the cutoff age? How are adults immune to the effects of electricity? In scientific fact, carcinogens have no intellectual capacity to make any decisions about what group of humans or animals to target. Carcinogens cause cancer in children and adults at about equal rates, but data can be selected and manipulated to "prove" otherwise.

Statistical Abuse

Physicians have an insatiable appetite for statistics in their research. I have yet to read an article submitted by a physician who has properly applied the principles of statistics. Many of you have heard the expression that "statistics can lie". Statistics can not lie. Researchers who use statistics can lie. It is easy to pick subjects such that the statistical analysis "proves" one's point. The relationship between childhood leukemia and power lines is a good example of how this can be done. A principle that physicians consistently fail to recognize is that if a researcher needs statistics to prove his point, then that researcher has no point to make. Please remember this statement whenever you come across stories about cancer.

Here is a six step procedure you can follow to statistically "prove" a link between cancer and power lines, or any object you choose. First, go to a place where the incidence of leukemia in children is relatively high. This data can be obtained from hospital records. Second, of the children who develop leukemia, use in your data base only those cases in which children live near power lines. Third, locate a place far away from any major power lines where the incidence of leukemia in children is low. Fourth, determine the number of children living in a square mile area in both locations and select 10,000 children living in the area around power lines and 10,000 children living in the area without power lines. Fifth, count the number of leukemia cases in each area. There might be 17 cases of leukemia in the children living around power lines and 10 cases in those children not living around power lines. Sixth, divide 17 by 10 to arrive at the 1.7 figure. You may need to ignore another 5,000 or more children who lived around power lines who never developed leukemia in order to stack the numbers in your favor. Your statistical calculations "proved" your point that living near power lines causes leukemia in children. If all else fails, make up your own numbers. It is highly unlikely that anyone will challenge your report effectively. By being the first to publish this fraudulent work, you will place doubts and fears into everyone's minds for a long time as you build a reputation.

Timely Press Releases

There is still another tactic physicians like to use. You all have heard news stories about physicians reporting "promising new leads in fighting cancer". Of course, further work in this area must be done to be certain of the results of the preliminary study. This is a technique they use to sucker you and politicians in to continuing to finance them. If you read the history of "cures" for cancer, you will find a change in the definition of what the researchers mean by a "cure". Earlier this cure was most often defined as extending the patient's life free of symptoms for perhaps 5 years. In more recent articles, the "cure" has been redefined as extending the patient's life by as little as 6 months. This time span varies based upon the statistical analysis of results, by the treatment strategy used and the type of cancer. For example, if using drug X extends the life of patients by 1.5 years over patients who do not receive this drug, then the physicians will conclude with statistical certainty that drug X is a "promising new lead" in treating that cancer. In reality, the cancer is never cured.

How they arrive at this extension of life expectancy is simple. Cancer patients are divided into two groups. One group does not get the new treatment. The second group gets the new treatment. If members of the second group live any longer than the untreated group, then physicians announce to the public that they have a promising new lead in fighting the cancer. Further, they feel that they can report with certainty how much longer the patient will live due to the new treatment. The reality is that <u>none</u> of the patients in the group who received the treatment are cured of the disease. The physicians merely redefined the definition of success.

In an example of this tactic, women with breast cancer were experimented on using a new treatment that involved the chemical bonding of an FDA approved cancer drug to a group of proteins called monoclonal antibodies. Monoclonal antibodies are suppose to selectively seek out a given type of cancer. The physicians removed the bone marrow from each one of the first group of women and treated it with the breast cancer cell monoclonal antibody with the cancer drug attached. The marrow was then replaced into each woman. The other group did not receive this treatment. Women in the treated group lived only an average of about one month longer than the women in the untreated group. None of the women in either group was free of breast cancer for more than about three years. All of the women in this study eventually died of breast cancer. The physicians claimed this treatment was a success. They wrote nothing of the quality of life experienced by the women in either treatment group. An extension of life is important, but it is not a cure for cancer.

The truth is that physicians are no more successful today in curing cancer that they were 10 or more years ago. They are only redefining the success rate to last for fewer years in an attempt to appease the public. Even though some patients go into remission for as long as 20 years, these cases are few and they do not accurately indicate that physicians have improved their success in fighting cancer.

Disqualified

These research projects are fueled by the ego. If any physician could solve the riddles of any part of the trilogy of cancer, it is certain that he or she would win the Nobel Prize for medicine. This individual also will become a historical figure, and of course be set for life financially.

Physicians without a Ph.D. degree are ill qualified to lead cancer research programs. They have consistently demonstrated their ineptness at understanding this disease. Their training and education is inadequate to enable them to conduct scientific research. Unfortunately, cancer researchers with other educational backgrounds are not much better scientists than physicians, as you shall learn.

CHAPTER 4

MOLECULAR BIOLOGISTS

A second group of researchers who would appear to have the appropiate credentials to study cancer are molecular biologists. They examine the genetic code to locate genes, a process termed mapping. Genes are responsible for making each of us different and at the same time somewhat the same from a biochemical perspective. The major concept in vogue with molecular biologists today is the oncogene theory. The prefix onco refers to cancer. Simply put, oncogenes are the genetic material that code for the formation of cancer, according to the molecular biologists. For different types of cancer there are supposedly separate corresponding oncogenes. A normal cell is transformed into a cancerous cell because an oncogene is turned on. Their strategy for the cure of cancer is to locate the oncogene which is in the "on" position and use some sort of agent to turn that gene "off".

The Scam

After more than ten years of promoting and supposedly working on this oncogene theory, the molecular biologists have failed to prove that oncogenes even exist.

Never mind that their isolation could not actually contribute to understanding the mechanism or cure. They are experts, though, at public relations and hype. In the news media reports about cancer, you will often hear such phrases as "Researchers have found a new gene. This gene offers exciting new leads to the understanding of cancer, or to a cure for cancer." But, this is where the news story stops. There is never even a brief explanation of how the isolation of the gene in question will lead to curing cancer. The truth is that there is no way. Molecular biologists rely on you to fill in the gaps of their logic. You are suppose to believe that locating a gene can cure cancer, and to keep your money flowing to them.

Although molecular biologists take your money with the promise of curing cancer, it is spent pursuing work with no real relationship to this disease. They are preying upon your emotions and good faith to get money to support pet projects on gene mapping. They tag on the statement that this work will increase the understanding of cancer because they know that if they were to ask for money to simply isolate genes for the sake of knowing their location in the huge DNA molecule, you probably would not fund them. By linking their work to cancer under generally vague and faulty premises, they solicit support for their "progress" against this disease.

Let me present a typical example of how molecular biologists work their scam. There is a group of molecular biologists employed by Bionetics Research, Inc. at Fort Dietrich in Frederick, Maryland who are working with the bacterium E. coli. This bacterium can not cause cancer nor does it possess any chemical that could be used to treat cancer. It is the bacterium the news media frequently cites in relation to genetic engineering experiments. Under the microscope, naturally occuring, or wild-type, E. coli is brown. This group of molecular biologists engineered a colony of this bacterium to appear blue. They then engineered another colony to be white. When I left this research facility, they were working on a red colony. You are paying for molecular biologists to make multicolored bacteria. None of the colored bacteria contains any substance that can cure cancer. Nor can changing the color of bacteria suggest a mechanism for the formation of cancer. These experiments are a mindless exercise in genetic manipulation.

Origin of Molecular Biology

In order to understand how molecular biologists have been getting away with their scam for over 10 years, a review of the history of molecular biology is helpful. Back about 15 or so years ago, there was no such field as molecular biology. The term used to describe individuals interested in genes was either geneticist or cell biologist. Both types of researchers were employed to examine cancer. Neither was able to offer any mechanism for cancer formation, or any cure. These people were losing status, which meant that they were losing jobs and prestige. In an effort to regain some measure of dominance in the general area of medical research, a splinter fraction formed the discipline of molecular biology.

About 10 years ago, molecular biologists were successful in isolating the genes for human insulin and then placing those genes into E. coli so the bacterium could manufacture human insulin. Once this bacterium made the insulin, it became a relatively simple process for scientists to harvest this protein. This was heralded at the time as a major breakthrough in both understanding and curing diabetes. Molecular biologists have been riding the crest of this discovery ever since, claiming that they can apply the same principles to finding a cure for cancer. Having a bacterium make human insulin cheaply and rapidly certainly is a great benefit to diabetics, but this is only a treatment for this disease and not a cure. Cancer is a much more complex disease than diabetes and certainly not amenable to the same strategy. As you will read, their claims can never be realized. It was the simultaneous failures of scientists in other fields of medical research to answer fundamental questions about cancer coupled with the serendipitous discovery of the genes for human insulin that provided the impetus for molecular biologists to gain the power they now enjoy in cancer research.

The Molecular Biologist

Let us examine more closely what the term molecular biology actually means. Molecular refers to molecules. Molecules are chemicals. Biology refers to the study of life. Therefore, molecular biologists claim to study the chemicals of life. In reality, they have a very limited understanding of chemistry. But they do find it easy to maintain their pretense in well funded research programs, when there is no one in a position to successfully challenge their work. Molecular biologists concentrate almost exclusively on the DNA molecule. They refuse to even consider any concept that does not pretend to study DNA. A molecular biologists once literally screamed at me that she refused to believe that anything of importance to cancer research occurs away from the DNA. This arrogant intimidation is typical of the molecular biologist's approach to dealing with the more technically astute scientists who don't buy into the oncogene theory hoax. Unfortunately, this tactic is generally quite effective because molecular biologists direct many of the cancer research programs today.

In reality, no one can cure cancer by merely locating genes. Molecular biologists do not truly understand the very discipline they lay claim to. Locating a gene is not equivalent to knowing how that gene works, or is expressed. For example, both muscular dystrophy and juvenile diabetes are genetically determined. No molecular biologists to date has been able to use this knowledge to cure either of these diseases by using genetic engineering techniques to correct the faulty genes. Likewise, no cancer researcher to date has been able to cure cancer with the knowledge that genetic alterations have occured. The oncogene hoax has provided false hope. Molecular biologists will never cure cancer.

A Very Short Course in MB

To better understand how molecular biologists operate, you need a quick and easy course in molecular biology. Inside each cell in our bodies there is the full complement of genetic material or DNA (DeoxyriboNucleic Acid). Each liver cell has exactly the same DNA as each heart cell and nerve cell. The reason a given cell is one type as opposed to another is that different sets of genes were expressed.

DNA codes for RNA (RiboNucleic Acid). Through a complex series of events, the RNA becomes transcribed into proteins. To sum up the process, DNA codes for RNA, and RNA codes for proteins. The DNA molecule is made from four nucleic acids strung together in various sequences, which correspond to genes. It may be helpful for you to visualize the DNA molecule as a string of beads made of four colors: red, blue, green and yellow. These beads are strung together in just about any sequence you can imagine. Certain segments in this string correspond to individual genes. Any given gene can be comprised of about 10,000 nucleic acids, or beads.

The High Price of Genes

Chemists developed a procedure to isolate DNA from tissues. I have performed this procedure several dozen times. It is a rather long process described in about 10 typewritten pages. The isolation of DNA from tissues takes at least an 8 hour day to complete. The chemicals required for this procedure are relatively inexpensive. For about $100 you can purchase enough materials to perform 200 to 300 extractions of DNA. This phase of working with genes is inexpensive and relatively easy to do.

Costs for performing molecular biology experiments rapidly escalate when they are done to find a particular gene. When the DNA is harvested from a sacrificed (i.e., killed) animal, all of the genes are present in one relatively large mass. Chemists have developed methods to separate this large mass into fragments. Biochemists have developed methods to cut these fragments into smaller pieces, which are then separated. One of these smaller pieces will contain the gene of interest to the molecular biologist. The technique used to separate DNA fragments is called gel electrophoresis.

Electrophoresis is based upon simple scientific principles. In much the same way as you would make a gelatin mold, the gel for electrophoresis is made in a tube about the size of a standard sipping straw. A positive electrode is applied to one end of the tube and a negative electrode is applied to the other end of the tube. The DNA sample is placed at the negative end of the tube. A direct current of 500 to 1000 volts is applied. Fragments of DNA migrate within the tube at a rate based on their size. The smallest fragments travel the farthest and the largest fragments move the least distance. This migration of the DNA sample is allowed to continue overnight. When the migration is complete, alternating light and dark bands appear along the length of the tube. The bands are visualized typically by using radioactive phosphorus-32 (P-32) or a dye named Coomassie Brilliant Blue G. Dark bands designate the location of DNA fragments, while the light bands indicate no DNA fragment at that spot.

The technique of gel electrophoresis is simple. All you have to do is follow the instructions provided by the suppliers of the equipment and chemical reagents used. Surprisingly, the equipment is the smallest expense for this type of experiment. The chemical reagents can cost thousands of dollars for one set of experiments. When I worked at Bionetics Research, Inc., I happened to see the invoice for one of the individuals who was to conduct an experiment in molecular biology. About six chemical reagents were ordered. The price for these reagents came to just over $4,000. This was only one experiment of a number planned. This experiment was allowed to be conducted even though the reasons for doing it were severely criticized by an independent committee. The committe concluded that this series of experiments, designed to find oncogenes, would not add any significant information to the understanding of the mechanism of cancer.

Let us examine the costs associated with starting up and operating your own molecular biology laboratory. Again we are assuming that the building has been already acquired. The molecular biology laboratory can not be in the same room as your mouse. Toxic chemicals, including radioactive compounds, will be used in your molecular biology experiments; the mouse can not be exposed to these. While most molecular biologists shun using a chemical fume hood, I strongly suggest you buy one, for safety reasons, to prepare the reagents in. Chemical fume hoods cost on average $8,000. The hood will have to be connected to ducts leading to an exhast fan

on the roof. The duct work and fan cost about $3,0 ̲ ̲ ̲ ̲ ̲ ̲
parts and installation. The equipment you will need to
purchase is listed in Table 4-1.

Table 4-1. Equipment List to Separate DNA Fragments

Item	Price
Chemical fume hood	$8,000
Fan and duct work	$3,000
Centrifuge	$30,000
Electrophoresis apparatus	$2,500
Top-loading balance	$1,500
Thermostatically controlled water bath	$1,800
Tissue homogenizer	$1,000
Chemical and storage cabinets	$1,500
Counters with drawers	$1,000
pH meter	$175
Hot-plate with magnetic stirrer	$150
Radiation shield	$200
X-ray film developer	$300
Glassware	$1,000
Miscellaneous	$500
Total	**$41,625**

When you add the costs for renovations and installation to the equipment list, you can have a basic molecular biology laboratory for roughly $53,000. You still have not purchased any chemicals at this point. The price of the chemicals you will need is staggering. In order to cut the DNA molecule into pieces, a group of enzymes called restriction endonuclease enzymes are used. Restriction enzymes are isolated from a number of sources, including rattle snake venom, bovine pancreas (the term used by researchers for cow is bovine) and various yeasts and bacteria.On average, restriction endonucleases cost about $25 in the minimum amount you can buy. Because each endonuclease cleaves the DNA molecule at a different site, you will have to purchase several. If you purchase 10 different enzymes, it will cost at least $250 and typically much more. Molecular biologists routinely buy many more than just 10 enzymes. You'll find it easy to spend well over $1,000 just on enzymes.

In order to image the DNA fragments generated by the enzymes, you have to use another set of enzymes that "read" those fragments and then make new DNA fragments. During this process, radioactively labelled phosphorus, P-32, is used. In the reaction flask there sits a mixture of various sized DNA fragments called the digest. To this mixture a set of enzymes called polymerases is added along with P-32 in the form of phosphate and the four nucleic acids. The flask is placed into the thermostatically set water bath and the mixture is gently shaken overnight. In the morning, you have the new DNA fragments which have incorporated the radioactive P-32. This mixture is then placed onto the columns of gel you prepared the day before. The current is hooked up to the columns of gel and left on overnight. The following day, the columns are carefully removed from the tubes and placed on the x-ray developer for at least 24 hours. When the film is developed, you will see dark bands where the DNA fragments are located. This procedure is relatively simple if you follow the instructions.

To summarize the entire procedure, it takes a full 8 hour day to extract the DNA from the animal; a second day to prepare the gel and the DNA digest; a third day for the polymerases to make new DNA fragments; a fourth day to separate the DNA fragments; and a fifth day to develop the x-ray film. One experiment in molecular biology takes a full week to complete. This one experiment typically costs about $1,000 for chemicals alone.

When molecular biologists describe their work, they imply that they know with reasonable certainty where to locate the gene they are after, usually any one of a number of oncogenes. The phrase they like to use is "active site directed" agents which they claim target a gene of interest. This is hardly true. No molecular biologist has a clue as to where most genes are located in the DNA molecule at the time an experiment is started. Consequently, they have to use an array of restriction enzymes to generate thousands of DNA fragments. Experience has shown that the smaller fragments probably do not contain an entire gene, so these fragments can be ignored in the search for an oncogene. To find the gene in question, molecular biologists have to conduct hundreds and even thousands of experiments similar to the one described above. If we take as an average cost of $1,000 per experiment for chemicals alone and multiply this figure by 1000 experiments, a typical number, it will cost $1,000,000 to find any one specific gene. The price tag to find the specific gene is often about $150 million.

Locating genes is often a subjective process. Ideally, each of the columns of gel are properly aligned at one end. Molecular biologists then compare the location of each band. If there is a band at one location on one gel column but that band is absent from a second column, then molecular biologists believe they have found a new gene. When molecular biologists search for oncogenes, they take DNA from normal tissue and DNA from cancerous tissue. The two sets of DNA are supposedly treated identically in the gel separation. Again, they compare columns to find differences in the band patterns. If they find a band at one location in the column containing DNA from cancerous tissue that is absent from normal tissue, they claim an oncogene was found. There are, however, other explanations for the differences in band patterns from the two sets of DNA.

Designed Experiments
Faked Results

One simple and straightforward explanation is that the results are fake. Remember, the molecular biologist is interested in demonstrating that oncogenes exist because his job and reputation are on the line. With no oncogenes, there is no job. After the DNA digests are developed to reveal the fragmentation patterns, there may not be any real difference between the cancerous DNA and the normal tissue. This is an easy problem to resolve. The columns of gel can be realigned such that differences in the fragmentation pattern can be seen. The original development of the gel is withheld from the view of others. The molecular biologist presents only the patterns he chooses to support the oncogene theory.

Ghost Bands

Another tactic molecular biologists use is to boldly state either that a band is present when in fact it is absent, or to state that a band is absent when in fact you can see the band on the developed film. I have attended several lectures and seminars given by molecular biologists to summarize their work. I have witnessed this tactic on every occasion. Virtually every chemist in the room also picks up the discrepancies in the data. The typical response given by the molecular biologist when his data is challenged is that "I have better slides back at the lab". We chemists wonder then why he didn't bring those "better slides" to the seminar. The real answer is that no such slides exist. He brought the best or most convincing set of data he had. This data does not support the oncogene theory, yet he tried to ramrod it through the scientific community.

A Bad Prep

A third explanation for the differences in the DNA fragmentation patterns can be found in the way the tissue sample was prepared. This preparation includes making up the reagents to be used in the experiment. Every supplier of the enzymes used for molecular biology warns of the possibility that artifacts, or false results, can occur if the detailed set of instructions are not followed properly. If the tissue sample is allowed to stay around too long, deterioration of the DNA results, which shows up as differences in fragmentation patterns. In the natural course of the biochemistry of the cell, normal or cancerous, a process called methylation of the DNA occurs. This methylation can lead to differences in DNA fragmentation patterns, which also will be seen when the film is developed. There are a number of explanations for differences in the fragmentation patterns of DNA which are due to chemical reasons and not to the existence of oncogenes. None of these possibilities are ever addressed by molecular biologists because they do not understand chemistry.

Culture Shock

Separation of DNA fragments is one half of the set of experiments to locate a gene. There has to be a method to verify that you have the gene. This test usually involves taking the DNA fragments and inserting them into the bacterium E. coli.

When you work with bacteria, you are working with tissue culture. This means that another laboratory has to be set up for these types of experiments. A tissue culture laboratory will cost about another $32,000 for equipment and installation. For this laboratory you need another type of hood, a laminar flow biological safety cabinet costing $8,000, designed to keep the work area sterile. You also will need an incubator, another $8,000, to house the bacterial colonies in a temperature and humidity controlled environment. The duct work for the laminar flow hood must be separate from the chemical fume hood. Plan on a $4,000 installation charge for the hood and the incubator.

There are two methods to measure the expression of genes. Both require the purchase of a spectrophotometer. This instrument is used to determine gene expression by measuring either an increase in proteins or an increase in RNA. These instruments have a wide range of prices, but expect to pay $10,000 for one that meets your needs. This instrument can be placed in the same room as the electrophoresis apparatus. You will have to buy accessories for this instrument for an additional $2,000. It is easy to understand how molecular biology work is the most expensive type of routinely conducted medical experiments. The cost of molecular biology experiments is summarize in Table 4-2.

Over $150,000,000 spent on a false claim that an oncogene was located. And they don't even exist! This total does not include the cost of removing radioactive wastes, or utilities and salaries. Nor does the price include obtaining a license to keep and use radioactive materials. I was quoted a price of $50,000 per year by the FDA for a license to house radioactive compounds.

Table 4-2. Summary of Costs for Molecular Biology
Experiments

Item	Price
DNA extraction reagents	$100
DNA separation apparatus and reagents	$53,000
Gene search	$150,000,000
Tissue culture laboratory including spectrophotometer	$32,000
Total	**$150,085,100**

Radioactive Waste

Radioactive waste is an unfortunate side effect of molecular biology research. It is difficult to estimate the cost to the quality of life and to the environment that this pollutant exacts. When molecular biologists describe their work, they never mention the quantity of radioactive compounds they use. An estimate can be made, though, be reviewing the sales figures of New England Nuclear, which manufactures most of the radioactive compounds used in all types of cancer research. By taking into consideration all of the cancer research facilities throughout the world, it is safe to estimate that tons of radioactive wastes are generated by cancer researchers each year. The overwhelming majority of this waste comes from molecular biologists.

It is alarming how cavalier molecular biologists can be about handling radioactive wastes. I worked at Oak Ridge National Laboratories in the Biology division, which was only a few hundred yards from where nuclear warheads were assembled. You would think that someone there would have some knowledge of how to store radioactive wastes until they were removed from the premises. This was not so. Radioactive wastes, both solid and liquid, were placed in a 55 gallon drum. This drum was in the hall of the building. There was no chemical safety hood near the drum to vent volatile radioactive fumes. There were no safety shields to prevent exposure to radiation. The drum just sat in the hall with a yellow tape on the floor around it. A similar situation existed at the University of Tennessee Veterinary Teaching College and at Bionetics Research, Inc. The only difference in procedure at Fort Dietrich was that solid and liquid radioactive wastes had to be separated. At this facility, the Army crushed the solid wastes into tight packages to be buried in dump sites.

A Total Waste

None of the work conducted by molecular biologists has any real impact on cancer. They use a weak connection to cancer research to seduce you and politicians into giving them billions of dollars to continue their simplistic and worthless gene mapping projects with the real goal of having their egos massaged.

CHAPTER 5

BIOCHEMISTS AND CHEMISTS

Cancer researchers for several decades have understood that chemistry plays a vital role in the mechanism of cancer. Over thirty years ago, biochemists examined differences in protein content between normal and cancer cells as a means to unraveling this mechanism. No definitive patterns in protein content were consistently observed at that time. Consequently, biochemists were demoted unofficially to lesser roles in cancer research. Biochemists and chemists have, nevertheless, made major advances in understanding the mechanism of cancer.

Biochemists study the activities of molecules that are found in all life forms. These molecules include DNA and enzymes. They also are very interested in how different chemicals are handled in the body. Because it is generally agreed by all cancer researchers that chemical carcinogens must be acted upon by enzymes to initiate the carcinogenic process, biochemists seek to find which enzyme systems are responsible for this metabolism and to determine the chemical products, or metabolites, of this metabolism.

In the context of this book, I define chemists as organic chemists. These scientists enjoy combining various chemicals to form new compounds. They are the individuals who try to develop new anti-cancer drugs. In groups of cancer researchers, biochemists and chemists often work as a team. They are in constant communication with each other. Their areas of expertise overlap so much that it is difficult at times to determine who is the biochemist and who is the chemist.

Selectively Open-Minded

Biochemists and chemists tend to be the most open to new ideas in cancer research. They probably have done more to advance our understanding of cancer than physicians and molecular biologists combined. However, there are a few egotistical individuals in high positions who have their pet theories. Young, talented biochemists and chemists entering the field of cancer research are not allowed to challenge those theories. As a result, there is a stagnation in the development of new strategies in understanding the mechanism of cancer and discovering cures.

One of the generally agreed upon concepts in cancer research is that most chemical carcinogens are not the actual agents that initiate the cancer process. These chemicals must first be metabolized to form active metabolites. Metabolism in this context means an alteration in the chemical structure of the initial compound. Because this metabolic process occurs, it is sensible to examine how these chemicals are transformed.

When biochemists speak of metabolism, they are referring to specific enzymes or enzyme systems responsible for the alteration of chemicals. Based on experience with drug metabolism, biochemists know that the liver is the main organ where the metabolism of chemicals occurs. Using the livers of animals is one method biochemists have to examine how chemcial carcinogens are transformed in the body to initiate the process of cancer formation. Sometimes it is necessary to use the whole animal. To conduct their experiments, biochemists collect the urine and feces of the animal to analyze for any metabolites that may be found. The theory behind the biochemists' work is that by knowing which enzymes are involved in the transformation of chemicals into carcinogens, they can better understand the complex series of biochemical events required to initiate the cancer forming process. In having this knowledge, they feel confident in their ability to unravel the mechanism of cancer and to offer new strategies for a cure.

While biochemists and chemists are the best candidates for unraveling the mechanism of cancer, an obvious question is why they have not done so. The answer is the same as for physicians and molecular biologists. Ego! Biochemist and chemists also have their pet theories, including which enzymes transform the chemcials into carcinogens. In order to be able to work with biochemists in positions of authority, subordinates must use only carefully defined enzyme systems. Using any other enzyme system is strictly forbidden because this runs contrary to their theory. I discovered by experience that they even frown upon using organs other than the liver to study the metabolism of chemical carcinogens.

When I started working at Bionetics Research, Inc., I was interested in continuing work on the mechanism of lung carcinogenesis. Having worked with human lung cancer cell lines in a previous position, I wanted to use the whole animal to apply information I had obtained in tissue culture experiments. I thought my research plans were clear to everyone in my department. I emphatically stated my interest in working with any organ other than the liver. I repeated my research focus by stating that I did not want to work with the liver of any animal because too much work had been done with that organ without producing an understanding of how lung cancer forms. Inspite of my clearly stated research interest, the word I received through the cancer research grapevine was that I was looking at the metabolism of chemical carcinogens using the liver of rats. This was a not too subtle message that I was going against established policy and that I should change the focus of my research. Even with this pressure to conform with an unwritten but established policy, I continued to examine the roles of other enzyme systems in organs besides the liver, and was branded a troublemaker.

What Biochemist Do, and Don't

Because biochemists can not use human subjects to study the metabolism of chemical carcinogens, they use animal models to study this process. Most of this work has been conducted using either mouse or rat livers. Within the liver there is a system of enzymes called the mixed function oxygenases, or P-450's, that are responsible for the metabolism of chemicals. These chemicals can be drugs, carcinogens and even the hormones manufactured in our bodies. Study of the liver and the P-450 enzyme system was an obvious choice for the biochemists. A relatively simple laboratory procedure to isolate P-450's from the livers of animals was developed.

One of the interesting aspects of the metabolism of chemical carcinogens is that very active metabolites are produced that bond to the DNA molecule. Biochemists have worked out experimental procedures by which they can precisely determine where the bonding occurs and quantitate the amount of chemical bound. When this chemical bonding occurs, it can lead to the subsequent alteration of genes. This is the point where the molecular biologists enter to try to determine which genes become involved. Some of this work by biochemists led to the development of the oncogene theory in cancer research.

It would appear that all the pieces of the puzzle are together now to explain the mechanism of cancer when you combine the findings of molecular biologists with those of biochemists. The supposed series of events for the initiation of cancer is as follows. A person is exposed to a chemical carcinogen. The liver metabolizes this chemical to produce a metabolite that bonds to DNA. This bonding causes the oncogene to become expressed and the person develops cancer.

Interesting theory, but incomplete. The process of carcinogenesis (cancer formation) is more complex than most molecular biologists and biochemists understand. There is a pattern of cancer formation in the various organs. Not all chemical carcinogens cause cancer of the liver even though the compound may be metabolized in that organ. One class of chemical carcinogens are called nitrosamines. These chemcials are found in tobacco products, cutting oils, cosmetics and some foods. These chemcials can be considered to be found virtually every-where in our environment. It is logical, therefore, for biochemists and chemists to study how the body handles these compounds. The metabolism of nitrosamines is an important issue in cancer research.

The fact that a chemical carcinogen can be metabolized in the liver without causing liver cancer can be illustrated with diethylnitrosamine. Cancer researchers refer to this compound as simply DEN. Some DEN is found in all tobacco products. It is one of the main chemical carcinogens in cigarette and cigar smoke. Most cancer researchers agree that it is one of the most potent carcinogens known. Cancer has been produced in all animal species dosed with this compound. However, there is a specific pattern to the organs in which cancer is observed. In most animal species it is a respiratory tract carcinogen. This is to say that tumors are consistently found in the nasal cavity, the bronchioles and down into the lobes of the lung. This is exactly the same pattern of tumor incidence found in humans who smoke. In some animal species there have been reports of tumors in the liver also. Tumors in the kidneys, pancreas, intestine, stomach or any other organ are rarely found. Due to this consistent pattern of respiratory tract tumors, DEN is considered a lung carcinogen.

Progress by Decrees

It has been decreed by some biochemist that in order to understand the metabolism of chemical carcinogens the liver must be used. Work with another organ, such as the lung in the case of DEN, will be severely challenged. Further, it also has been decreed that the P-450's are the only enzymes of importance in cancer research. There are many other enzyme systems that continuously operate, but if you explore these alternative pathways, then it will be nearly impossible to get your work published. I can make these statements with certainty because I have experienced this suppression, and I have heard from other cancer researchers who have similarly suffered. This heavy handed, narrowminded policy is enforced by the biochemists with clout.

When I first began in cancer research, my supervisor was interested in examining the metabolism of DEN in both human lung cancer cell lines and in the Syrian golden hamster as an animal model for human lung carcinogenesis. I was fortunate in being able to work with a particular cell line that was very interesting for its lack of P-450 enzymes. Yet, this cell line could metabolize DEN at a rate about 200 times faster than the standard test system, rat liver. It was clear that some other enzyme had to be responsible. The question was which one. I made a model of DEN and observed some similarities between this chemical and a naturally occurring chemical our bodies produce and use everyday, named serotonin. This compound is metabolized by another enzyme system called monoamine oxidase, or MAO for short. But I was not the first researcher to make this observation.

A few years before I began my work, a pair of researchers named Lake and Cottrell published a series of papers on the metabolism of DEN and other nitrosamines. They demonstrated in these papers that MAO metabolized DEN. Their report set off a controversy among biochemists. Many refused to believe their results because these researchers used an enzyme system other than the P-450's. Some of this criticism was published; some biochemists expressed the same criticism to me personally. I could find no fault with either the reasoning or the methodology used by Lake and Cottrell.

Unknown to me at the time, there was a feud raging between Lake and Cottrell on one side and many other biochemists in cancer research on the other. To settle this dispute over whether or not MAO metabolizes DEN, a very prominent biochemist published an article. In this article he decreed that the controversy was over. MAO does not metabolize DEN and he presented his results to prove his statement. I examined his data carefully. This critic had no idea of how to use his own instruments. He used two different pieces of equipment, neither one of which could determine precisely what series of events occurred during the monitoring of the DEN metabolism.

First, the critic used an oxygen electrode to monitor the metabolism of DEN in the presence of MAO. This electrode only detects the rate at which oxygen is being used up in the reaction flask. The electrode can not determine how the oxygen is disappearing nor to what type of chemical the oxygen becomes bound. To conduct the experiment, MAO and a chemical that this enzyme will act on are added to the reaction solution along with the oxygen electrode. The electrode is turned on and the results are recorded automatically. To measure DEN metabolism, DEN is added to the above reaction mixture and the results recorded again. Now there are two chemicals in the reaction mixture. The critic alleged that if DEN inhibits the metabolism of the other chemical, then you should observe a drop in the rate of metabolism by seeing less oxygen used up. However, this may not be true because the electrode only measures a drop in oxygen concentration. The oxygen may bond to the first chemical and to DEN at the same rate as it bound to just the first chemical alone. Not observing a change in the rate of oxygen usage does not necessarily mean that the enzyme did not act upon DEN, although this is what the critic claimed.

In a second experiment, the critic used an instrument called a spectrophotometer to measure the rate of metabolism of the first chemical by MAO. Again, only this chemical was used to obtain a standard rate of metabolism. DEN was then added and the rate again monitored. He reported no change in the rate of metabolism of the chemical. I tried to reproduce the exact technique he used. It was impossible to stabilize the instrument because the presence of both compounds caused it to act extremely erratically. There was no possible way to measure the rate of either compound. When I asked another biochemist why I could not make this instrument stabilize, he confirmed that it was due to the interference caused by the presence of both chemicals in the reaction mixture. It is interesting to note that the critic did not include a reprint of the spectrophotometric data in his article. This critic also did not actually refute the results presented by Lake and Cottrell, he just ignored them. Nevertheless, his conclusions were generally accepted without challenge because they supported the conventional beliefs.

I tried to reproduce the general results reported by Lake and Cottrell using the same test method they had published. I also used the same type of instrument the critic used in a companion experiment. To explore the hypothesis that DEN is metabolized by MAO, I used four different model systems. These were a commercially available purified MAO preparation, isolated crude enzyme preparations from human lung cancer cell lines, intact human lung cancer cell lines, and the intact Syrian golden hamster. I measured the metabolism of DEN by MAO using four different methods. These were the colorimetric method described by Lake and Cottrell, high performance liquid chromatography (HPLC), the same type of spectrophotometer used by the critic, and a pharmacological technique with the whole animal. In every experiment, I was able to measure DEN metabolism by MAO. My experimental results agreed with Lake and Cottrell, who had published their work about 5 years before I started mine. I was warned by one of the biochemists I worked with at the time not to publish my results because I would meet the same criticisms that Lake and Cottrell had faced. Another biochemist on a review committee also issued a stern warning not to try to publish my results, but he was more threatening about my future career. Both of these biochemists knew the critic well and did not want to risk his anger. This is a common example of how these egotistical self proclaimed "gurus" repress young scientists with novel ideas. To date, I have not been able to publish my work with DEN and MAO. You will learn how important this enzyme is in understanding the mechanism of cancer when I present my theory later in this book.

Selective Insight

Outside of cancer research circles another fact about chemical carcinogenesis not generally known is that when an animal is dosed with a given carcinogen, tumors develop only in specific organs. The cancer does not spread throughout its body. For example, as stated above, DEN is a respiratory tract carcinogen, with no tumors observed in other organs. If the chemical structure of DEN is altered only slightly, as in dimethylnitrosamine (DMN), an entirely different pattern of tumor development is observed. Before I entered cancer research, no one could account for this phenomenon. Because I have a strong background in pharmacology, I was able to develop a theory to explain these results.

About six years ago as of this writing, I introduced to my supervisor a concept well documented in pharmacology called selective uptake. This term means that a cell type located in an organ will preferentially take in a drug over another cell type located even in the same organ. This is one of the basic principles in understanding how drugs are able to produce their therapeutic effects in the body. For example, if you have a headache or muscle ache and you take aspirin for the pain, the aspirin circulates in your blood stream until it comes into contact with the nerve cells that detect pain. Aspirin is taken up by these nerve cells to help alleviate the pain. Aspirin will not be found in any large quantities in the pancreas or lung cells because those cell types do not have the ability to uptake aspirin.

It was obvious, to me, that chemical carcinogens behave according to the same principles as drug molecules. I designed and conducted a series of experiments to demonstrate the process of selective uptake of carcinogens by the target organs in which tumors develop. I was able to publish some of this work. Since the time I presented this concept at a cancer conference in Atlanta, other researchers have published similar results. My work is never mentioned in their articles.

Getting this concept accepted by cancer researchers was not easy because of an organic chemist who decreed to the world more than 20 years ago that chemical carcinogens enter all cells of the body at the same rate, and the same concentration of carcinogen is achieved throughout the body. This chemist clearly overstepped the boundaries of his expertise and his field, but he set a precedent in cancer research that remained unchallenged for years. I never saw his data nor read his work. Other cancer researchers described his work and conclusions to me. These researchers did so to alert me to the obstacles I would face in having the pharmacological concept of selective uptake introduced into cancer research.

The work with MAO and on the selective uptake process are just two examples of how the egos of influential biochemists and chemists dictate the path of cancer research. It would not be so bad if they were correct in their theories. But neither the biochemists nor the chemists have a fundamental grasp of the very subject they claim to have expertise in, much like the physicians and molecular biologists. Unless individuals like the ones described give their stamp of approval to new concepts, supporting articles will not get published. Further, those who do not tow the official party line risk their careers.

What Chemists Will, and Won't

The same heavy handed policy exists among the chemists who are working on a cure for cancer. Their theories also are outdated and not substantiated by the facts. Chemists are channelling their efforts into two main strategies for curing cancer. One group of chemists is trying to find possible anti-cancer drugs in plants and animals. The other group of chemists is synthesizing compounds in the laboratory, hoping to discover anti-cancer drugs. The approach by these two groups are not mutually exclusive; there is actually a great deal of overlap.

The Natural Approach - A Case Study

Obtaining pharmaceutical compounds from plants and animals is a branch of science called pharmacognosy. It is an old practice dating back to the ancient Egyptians. They extracted a form of aspirin for pain relief from the bark of willow trees. Digitalis, a heart medication extracted from the foxglove plant, was prescribed 200 years ago. Physicians in India used extracts from the rauwolfia plant to treat psychotic states and depression more than 500 years ago. The first forms of insulin to treat diabetics were isolated from calves. Penicillin was found in bread molds and prescribed as a very effective antibiotic. There are a number of drugs isolated from plants and animals that are used today to treat human diseases. Because of these successes, a popular belief exists that there might be some sort of plant, animal or microorganism that contains anti-cancer properties.

While working at Bionetics Research, Inc. I had attended a lecture given by an organic chemist from California whose project was the isolation of natural products from an animal called the sea cucumber, which is found in the waters of the Great Barrier Reef off the coast of Australia. His reason for examining the utility of the sea cucumber was simple. Sea cucumbers have been on this planet for millions of years without evidence of having evolved from their original state. They are a very slow moving creature with no defenses, such as a hard shell or the ability to burrow under the sand. These animals have no apparent enemies. This researcher theorized that the sea cucumber must produce some sort of toxic chemical(s) that are natural deterents to ward off attack. He noted further that there have been no recorded incidence of cancer in this animal. He also pointed out that these animals were free of infection by microorgansms. He concluded, therefore, that the sea cucumber must manufacture some chemicals with a potential as anti-cancer drugs and possibly even antibiotics.

This researcher obtained a $5 million government grant to fund his research project, which required him to go to Australia to harvest sea cucumbers in order to extract chemicals to be tried as potential anti-cancer drugs. This was an enormous undertaking. There are thousands of chemicals manufactured by any given plant or animal that are necessary for that organism to survive. From these thousands of chemicals only a few, perhaps three or four, have any potential as anti-cancer drugs. To search for these needles in the sea cucumber "haystack", he needed facilities to house his laboratory and a large research staff.

As I recall, he hired about 7 Ph.D. chemists and about 10 technicians for the project. The average salary for a Ph.D. at that time was about $40,000 per year and for a technician about $25,000 per year. He likely took about $75,000 per year salary for himself. He paid a total of $530,000 just for his technical staff. However, there had to be other support personnel, such as secretaries and maintenance persons, for an additional cost of about

$100,000 per year. This researcher was paying approximately $700,000 per year in salaries alone.

From the pictures he showed of his facilities, he had elaborate equipment, called extractors, custom built for his needs. This is the apparatus that isolates various chemicals from the sea cucumbers. The price for this equipment is estimated at $200,000. After the compounds are extracted from the sea cucmber, they have to be separated and the chemical structures identified. The precise determination of chemical structure is done by spectroscopic identification. Expensive and delicate instruments must be used. Of these, the mass spectrophotometer (MS) costs $150,000. Another instrument which you may have heard of in relation to cancer treatments is the nuclear magnetic resonance (NMR) spectrophotometer. The instrument organic chemists use is slightly different from the one used in cancer treatment. However, this instrument costs about $200,000. Estimated costs are based on average instruments that vary depending on features ordered. There are other major pieces of equipment required such as computers, laboratory furniture, materials and supplies which come to about $200,000. Assuming that no money was spent to build the laboratory, this cancer researcher spent $750,000 for equipment and supplies. Of course these majors purchases will not have to be made every year, but there are associated costs in operating and maintaining this equipment that can approach $100,000 per year.

Harvesting sea cucumbers was another major cost associated with this research project. The researcher has his facility in southern California, but the animals are obtained only in the waters off Australia. He packed up his family and flew them along with some members of his staff to Australia for several weeks each year to harvest sea cucumbers. His family members helped him on the project. To harvest sea cucumbers, he had to hire a guide, a boat and diving equipment. These rentals can cost $1,000 per week. In addition, he had to provide lodging and meals which could be another $1,000 per week. Then there was the cost of shipping live sea cucumbers back to California. Based upon my experience selling salt water fish and other marine animals, no more than two sea cucumbers can be shipped in one container. Also, this cargo must arrive in less than 48 hours for the animals to remain alive. A dead sea cucumber will begin to decompose rather quickly, destroying the chemicals to be extracted. The cost to ship one container can be $100. I estimate that at least 10 sea cucumbers are needed for one extraction experiment. This means that 5 containers must be shipped at a cost of $500. If 8 staff and family members flew round trip to Australia at $1,000 apiece, then the total price tag for airplane tickets was $8,000. A two week stay for this research group would come to a total of $12,500, just to retrieve perhaps 10 sea cucumbers. This researcher reported that at least two trips a year were necessary to harvest sea cucumbers because the amounts of chemicals extracted from the animals were very small. He would have spent about $25,000 per year just to harvest 20 sea cucumbers.

This cancer researcher had an annual operating budget of at least $850,000. This included salaries, travel and operating expenses. When the price of the equipment is included in the first year's budget, this researcher spent $1.6 million to begin operation. The price we paid for this project would have been well worth it, if he had discovered novel anti-cancer drugs. He did show structures of several compounds that he claimed had anti-cancer drug potential. Those structures were very complex. It would take organic chemists a long time to work out a reaction scheme to synthesize these compounds. Based on the experimental results presented by this researcher, however, none of those compounds offer any advantage over existing chemicals used as anti-cancer drugs, whether natural products or synthesized. But he got his dream of operating his own laboratory fulfilled. He gets to tour the world to discuss his project and thereby gain recognition from other cancer researchers. In short, his ego is fed and you continue to suffer and die from cancer.

Drugs and Poisons

Applying simple logic and appropriate knowledge, it is easy to understand why his project had to fail. This cancer researcher based his project on some assumptions about sea cucumbers that were never validly tested. It has been a romantic notion among many, including cancer researchers, that there is a chemical that will be found in some living organism which will cure cancer. I can apply this reasoning to the black widow spider in an effort to find an anti-cancer drug. A black widow spider has been on earth for millions of years. It also is an animal that has few natural enemies. There has never been a case of cancer reported in this species of spider. It can be concluded, therefore, that the black widow spider manufactures some chemical(s) that both ward off enemies, including microorganisms, and possess anti-cancer properties. Without much effort, I could isolate the toxin this spider produces to kill its prey. I could then test this compound in the tissue culture of any number of cancer cell varieties. I can guarantee you that the venom from the spider will kill virtually every type of cancer cell that I can test. But I can not consider this toxin to be an anti-cancer drug because it will not selectively kill only cancer cells. This spider produces one of the most toxic venoms known. If it were used in humans, it would kill the person along with the cancer. Most people would agree that death is not a viable option in curing cancer.

That the sea cucumber survived for millions of years on this planet with no reported cases of cancer can be explained adequately without assuming it manufactures potential anti-cancer drugs. The survival of this animal can be attributed to its habitat. It lives in relatively deep ocean waters of about 100 feet. Very few predatory species of marine animals, like sharks, crabs and starfish, dive to such depths. Consequently, this animal will have very few natural enemies. Lack of any microorganism infestation also is explained by the depth of the animal's habitat. Microorganisms are only found in waters at depths to about 50 feet. There is no infestation of the sea cucumber by viruses or other microorganisms because no such organisms are found in waters that deep. The complex chemicals found by the researcher in the sea cucumber may be produced simply to ward off predators by making the animal taste awful. Garden slugs, which use this same mechanism of defense, are not eaten by most species of birds because birds do not find the slug tasty, not because the slug is toxic. There are other valid explanations for the observations made by this researcher that do not rely on the implication that the sea cucumber produces natural products with anti-cancer potential. Also, most sea cucumbers never get regular medical exams, nor do black widow spiders.

While the chemicals extracted from the sea cucumber may kill cancer cells in tissue culture, it can not be concluded that these compounds offer any advantage over currently used chemicals. Nor does it prove that these compounds can have utility as anti-cancer drugs because just as the venom of the black widow spider kills many cell types, so too may the compounds found in the sea cucumber have the potential to be very toxic to humans. The rationale for using chemicals extracted from the sea cucumber is faulty. It is not based upon scientific principles and facts, only on romantic fantasy.

This sea cucumber project is typical of those you are funding, despite lack of any real basis in scientific fact or theory. There are many others. The head of each research project just keeps making grandiose claims of future success. Each of these projects is based on the same simplistic notion that if a plant or animal produces a chemical that either kills or wards off antagonists, then that compound is a potential anti-cancer drug. Of course toxic chemicals can kill cancer cells, but these chemicals indiscriminately kill normal cells as well, which means that you can die from the so-called cure. Anti-cancer drugs currently being used likewise lack the ability to differentiate between normal cells and cancer cells. This is the main reason there is no cure for cancer at this time.

Chemists can not rationally design or discover anti-cancer drugs because their backgrounds in pharmacology and physiology are inadequate for this task. Most are too stubborn and conceited to understand that this expertise is essential in the rational design of anti-cancer drugs. Their simplistic understanding of these subjects forces them to come up with bizarre chemicals that are poisonous and useless as anti-cancer drugs. Their well documented lack of progress in cancer research speaks for itself.

CHAPTER 6

THE RESEARCH LEADER

Working as a research scientist has many personal rewards other than salary. In fact, salaries for scientists are low when compared to lawyers, physicians, engineers and other professionals. My profession has an air of excitement due to its inherent exploratory nature. We are literally exploring new territories that no one has ever seen before.

In addition to this sense of adventure, there are satisfying creature comforts. We usually work in a comfortable environment. Discussions with colleagues are often interesting and stimulating. There is very little stress to perform, even for a researcher with limited investigative abilities. Work hours are usually flexible. It is easy to take time off for personal reasons, like going to the dentist or staying home to care for a sick child. This time off is compensated without taking a charge to sick leave or vacation. In general, it appears that research scientists lead a charmed life. Job satisfaction is, in fact, probably the main reason a person chooses to become a research scientist.

As nice a life as a researcher's may seem, it is very much better for those scientists who head a group or project. They have titles such as section head, group leader, director or vice-president of research and development. While their salaries do not compare to those of other professionals at a similarly high level, they enjoy more perks. Many of these indivduals have unlimited vacation time. They choose their own work hours. They attend several conventions a year in some of the most glamorous cities in our country and the world at either company or government expense. Sometimes they are called in by companies in other countries for consultation on a project. They get double pay for this service. Their employer pays them and the company or country they visit also pays them. Often the foreign host pays for their families to travel with them.

Visiting exciting and exotic places at double pay, on top of the routine perks, certainly is an inducement for scientists at that level to maintain their position and status. But financial considerations are pale in comparison to the voracious appetites of their egos. They thrive on believing that literally the whole world respects their mental abilities and achievements. They enjoy being singled out at conventions. They thrive on the power of spending millions of dollars per year to have their pet ideas explored. They gleefully enjoy hiring and firing other scientists at will. In short, they become intoxicated by power. But an intoxicated mind is a warped mind. This is one of the reasons why these alleged "distinquished" scientists fail in unravelling the mechanism of cancer or in discovering a cure. They are more interested in maintaining their power bases than in examining new ideas, unless they can claim those ideas as their own.

A Tangled Web

Maintaining this power base is a driving motivation for stealing ideas. The scientific community looks to the research leader for guidance in his field of expertise. He is under constant pressure to produce. This is particularly stressful because he probably never came up with an original idea before. He always played the political game in research. As an undergraduate he most likely dutifully did some project suggested by a faculty member in the department of his major, in order to earn a good letter of recommendation. He would then repeat this process in graduate school to earn an entry level position in a research organization. He would play the political game very carefully at this point. He befriends everyone and appears to be a "nice" guy. He basically sucks up to his supervisors while learning how to intimidate peers and subordinates. Meanwhile he listens to ideas discussed by other scientists at his level. Guess who is the first person to present some of these ideas to the supervisor? He gets credit instead of the scientist who originated the idea. The weasel gets the promotion and the originator of the idea looks for another job. It is by a series of such episodes that many of the so-called top cancer researchers in the world have risen to power.

As the weasel moves up the ranks, he soon notices a pattern of networking, that various cartels have been formed by cancer researchers. These cartels are similar in structure to drug cartels and just as competitive. Officially, all cancer reseachers will deny this political structure. They exist in cancer research, but they are cleverly disguised. Let us examine how they operate.

The Publishing Game

Suppose you want to publish your work in a scientific journal. Official policy for all of these journals is to publish articles in the appropriate field based upon the merit of the work. Theoretically, every scientist has the potential to publish in a given journal. Every molecular biologist has the possibility to publish in journals about molecular biology, but not in those founded for chemists.

Journals are suppose to be the forums scientists can use to share their knowledge with others. In reality, they are the most potent means by which heads of the cartels can exert their power and influence. They decide what articles are published and who is allowed to do so. As you probably guessed, only the articles that support the theoretical thrust of the leader, by researchers who pay him homage, get printed. It is a routine matter for them to find fault, on scientific grounds, with any of the other submissions. There is nothing a scientist can do to prevent this action by the cartel leader other than to try to publish in another journal.

The lacework of their operations is intricately and cleverly designed. A given cartel leader has a background in a specific field, molecular biology for example. However, he can also control what gets published in journals for other fields, like biochemistry and chemistry. Cartel bosses get this control by becoming editors for their journals, and taking positions as reviewer or assistant editor for journals in other fields. Through this network, one leader can control several journals.

As a young scientist just starting your career, you may not know to which cartel your supervisor belongs. But your supervisor knows. Your article will be submitted only to one of the journals participating in his cartel. Cartel members can not cross lines and expect to be published in journals controlled by members outside of their cartel. If the new researcher wants to play the game, because his ego needs feeding, then it is easy to use the cartel network to quickly advance his career.

In order to play this game, there are strict rules to be followed. First, never submit an article with your name alone on it. To do so is to risk severe retribution from your supervisor and any other researchers who feel they made a contribution to your work. Their feelings will be hurt, especially if they actually did contribute to your work. Retribution typically comes in the form of negative criticism of your work and excluding your name from their articles. If you still do not get the message, then you will be blacklisted, leading to the destruction of your career.

Second, there are a number of advantages to including as many names on your article as you can get away with. It will be easier to get the article published because you will have the endorsement of other cartel members. You will appear to them to be "a team player". By including their names, you open up the possibility of being included on their articles, increasing your publication list with no real effort. As your publication list increases, so does your reputation. It will be easier for you to get jobs and advancements.

A typical example of how networking within a cartel helps career advancement can illustrate my point. There can be four researchers within a cartel named Bob, Jill, Joe and Mike. Each one has an article to publish. If only one name is on their respective articles, then each researcher will have only one article on his publication list. But if each name is included on all four articles, then each member will have four articles to list. Very little, if any additional work has to be done by the researchers. I have witnessed one researcher make a casual phone call to another researcher. Their experiments were briefly discussed. For this reason alone, they included each other's name on the articles. High eschelon members in the cartels have huge phone bills.

Another way the cartel network is used to increase the publication list of members is to publish essentially the same data, but in other journals controlled by the cartel. If the cartel controls three journals, then the researcher can publish the same work, slightly re-worded, three times. If all four researchers get in on the deal, then only four experiments can produce up to 12 articles for each researcher within one year. To the uninitiated, the illusion is created that these are dedicated and prolific researchers. When they ask for more money to continue their work, of course they point to the "progress" they have made in fighting cancer.

Cartel networking is a way leaders extend their power base by building the credentials for themselves and cooperative subordinates. These subordinates are then placed in positions with universities, the federal government and pharmaceutical companies. The cartel leader places "his people" into these positions so that they can continue to work on supporting his theory.

When you hear or read that a study refutes the findings of a previous study, you may be the unwitting observer of two cartels engaged in a battle for territorial expansion, or a junior cartel member making a power move within his ranks. This point is discussed in more detail in the chapter on witch hunts. Pushing back the frontiers of cancer reseach is not their motivation, although the phrasing makes it appear so. The weapons used are poorly designed experiments and publications. Your tax dollars and contributions go to financing their arsenals in order for the egotists to battle each other in efforts to gain ultimate control over cancer research. Manipulation of data and subordinates are not the only battle tactics used by cancer researchers.

Of Bullies and Thieves

Theft of ideas in cancer research is common, and sometimes blatant. I attended a seminar once and offered some information on the pharmacological aspects of the project being discussed. Not even five minutes later, one of the other members in the group restated the very same idea I had presented as though he had just come up with it. He boldly tried to steal credit for my idea right in front of me. His supervisor bought his game. The rest of the group looked at the thief with disgust and commented to me later on this low class act.

The intellectual thieves eventually are called on to display their alleged talents for research. Of course they can not because they have been spoon fed ideas in the past. But they know that if they offer nothing new, they will face embarassment and disgrace within the scientific community. This exposure will mean the loss of the power and priveleges acquired over the years. Their solution is to surround themselves with talented minds, and to keep those minds subjugated so they will not challenge authority. It is these repressed scientists who supply the research leader with fresh ideas and maintain his exalted lifestyle.

Many of the leaders in cancer research do not originate the ideas they lay claim to, they merely present them. These ideas literally are stolen from subordinates who are in no position to effectively protest. These indentured servants are graduate students, post-doctoral research fellows and entry level scientists. These subordinates have two common characteristics: 1. They are young, and 2. They are powerless to act against their supervisor. The supervisor justifies the pirating of ideas by reminding his subordinates that they work in his laboratory and in cancer research at his pleasure. The threat of dismissal or withholding a good letter of recommendation are common terrorist tactics the so-called gifted cancer researchers use to great advantage. There is no way subordinates can protect themselves. If they object too vehemently to their ideas being stolen, they are blacklisted. In the relativley small, close-knit cancer research community, this means the end of a career.

There is nothing comparable to copyright or patent laws to protect these subordinates. They remain completely at the mercy of their supervisors. This situation urgently needs reform, and only secondarily for ethical considerations. Its fundamental failing is in inhibiting progress on real advances in cancer. But reform is not forthcoming from those who control this system and profit from it. You may be thinking that this sort of politics and theft of ideas routinely occurs in any business. This is true, and no more justified. In other industries morale, productivity and ultimately maybe even profitability suffer. But by keeping these tyrants in cancer research, millions of people continue to suffer and die.

Proud Failures

We have no voice in the policies established by the research power elite. We can not vote them out of their positions like we can a politician, but we can hold them accountable for not delivering the results we are paying them for. We constantly hear about "promising new leads" and "exciting new treatments" without getting real answers to any part of the trilogy. They betray the trust of cancer patients used as guinea pigs to justify continued project funding to budget committees. And to compound the indignity, they butcher the dead to record the results of their futile attempts, calling it an autopsy.

Personal and ethical failings aside, the research leaders must be periodically purged for nonperformance. They are hired and retained on their promise to make real headway against cancer, but somehow never meet the goals and objectives of which they arrogantly boast. The public has been seduced, and actually forced, into accepting their policies and strategies on cancer research without any real benefit. We have no say on the course of their research, but are kept hostage to their false claims of a cure for cancer being "just around the corner". They set policies on how we supposedly can prevent cancer, yet the incidence of all cancers is increasing at the rate of 30% per year. The exception is breast cancer in women, which is increasing at the rate of 37% per year. They choose which chemicals and activities, even sex, are carcinogenic. They give us theories on the formation of cancer, such as cancer being hereditary, that have no basis in fact. Because of their position and status, we believe them and continue funding them, oblivious to any alternatives. The reality of the situation is that none of them have made any significant contribution to truly understanding cancer. At best, they have provided us with only a few disjointed facts. They expect, and get, high praise for just presenting disconnected pieces of the cancer puzzle that they were actually hired to solve.

CHAPTER 7

UNIVERSITIES

So far I have discussed the reasons why neither the mechanism or any cure for cancer has been found due to the incompetent and egotistical individuals involved. They have failed miserably for decades, despite their claims to the contrary. It is intuitively obvious that this disease presents a more complex problem than perhaps any single research group can unravel. It would seem logical to put together research teams comprised of individuals from several scientific backgrounds to work on this disease. This grouping of personnel has been done at three major types of researching institutions: 1. universities, 2. federal facilities, and 3. private pharmaceutical companies. It has not proven to be particularly effective at any one.

Facilities, Staff, Patients, and Backstabbing

Almost all of the work in cancer is conducted at universities with associated medical schools and teaching hospitals. There are a number of reasons for this, the main one being that cancer research as it is conducted today is very expensive. Not only is the equipment costly, but the construction of facilities is expensive. These universities already have the facilities and easy access to hospital patients as a readily available supply for the collection of human cancer tissue samples, as well as test subjects for new treatments. They also maintain professional staffs from all scientific disciplines. Facilities, staff and availability of patients makes a university with a teaching hospital a compelling setting to conduct cancer research. Yet none of them can claim success.

University positions are staffed by the very same incompetent physicians, molecular biologists and chemists discussed in the previous three chapters. The researchers who head the projects at the universities are among the most egotistical anywhere. Each one has his or her own pet project. Their theories are based almost solely on speculation, not fact. The truth is that most of the data generated in cancer research refutes their theories. At these universities, not surprisingly, the head of research is often a physician, not a scientist. No one is hired unless he supports the thrust of the research goals at that university.

Cancer researchers at universities, as elsewhere, make a public show of mutual respect and cooperation. The reality is that each one is as competitive and eager as the other to lay claim to either unraveling the mechanism or discovering the cure. As a result, each researcher will do anything to undermine the work of a "colleague". This subterfuge includes stealing ideas from each other in an effort to publish promising ideas first and thereby taking credit. This practice was evidenced recently by the pair of researchers working at a California university who laid claim to the concept of oncogenes. There is evidence that this theory was first proposed by a colleague working in the laboratory of those researchers. However, the pair of researchers received the Nobel Prize for their alleged "discovery".

Principles of Plagiarism

As described previously, theft of ideas is routine in cancer research. Typically, the older cancer researchers bring in young scientists, pay them about minimum wage (this is what is called a post-doctoral research position) and milk these unfortunates for every idea and technique they can get. The head of the project takes all of the credit. This is one of the ways project heads rise to the top of their fields. But you pay a very high price for them to have their egos soothed. Not only do your tax dollars and medical expenses go to support their egos, you also pay with your lives. Having the wrong person in charge of a huge research project can only lead to failure.

There is a considerable difference between laying claim to an idea and the ability to follow through with that idea. Most often, only the person who developed the idea truly understands its application and intricacies. To have a so-called expert simply describe that idea to the public and to funding agencies is not adequate to ensure success.

For that pair of researchers who claim to have discovered oncogenes, I have a challenge. Once you have an oncogene, what are you going to do with it? Will you be able to explain what the mechanism of cancer is, or describe your strategy for a cure? I will bet that they can not adequately answer these questions because the real answer is that this theory is useless. The futility of this oncogene theory will be discussed in detail in a later chapter.

Most people think that cancer research has to be expensive, and as research is being conducted today, it is. But it does not have to be. I tried to get this message across to several universities throughout the country when I applied to teach chemistry. The notices for these positions stated that the ideal candidate must meet certain criteria including a well defined research project. I had some experience as a teaching assistant at Temple University and I had several years of work experience as well as a plan to continue my work on the development of anti-cancer drugs. In my resume I outlined the equipment and facilities I would need to do this work, and an approximate budget. Most of the schools to which I sent my resume never had the decency to ever send a rejection letter. This rudeness in academia is the norm, designed by weak minds to convey some level of superiority without performance from the buffoons responsible.

Of those schools that did have the courtesy to send a rejection letter, one response was especially interesting. The chairman of the search committee wrote that he was led to understand that I had accepted another job and, therefore, no longer wished to be considered for the position. As part of the resume, three letters of recommendation were required. I had never told any of these three individuals that I had accepted another job. In fact, I told each one of them that I very much wanted this position because it was at a pharmacy school that would have been an ideal location to continue my work. Apparently out of jealousy, professional or otherwise, one of these three individuals was deliberately trying to sabotage my career. I know who it wasn't.

One of the three who wrote a letter of recommendation on my behalf was my former advisor at Temple University School of Pharmacy. I kept in touch with him, as many of his former students do to keep him aware of our careers. On one occasion I visited him to discuss my plans to continue the work I had started as a graduate student on the development of anti-cancer drugs. He was impressed and tried to find a way to help. He came up with a possible solution that could benefit both me and Temple University. He recommended me for a teaching position to the dean of the school, which at that time needed teachers with a well defined research plan. I seemed to be an ideal candidate in my advisor's eyes. He had seen me interact with undergraduate and graduate students and knew I got along well with both. I interviewed twice with the dean for this position. The agreement was that I teach a couple of courses to the pharmacy students in exchange for a salary and some research funds. After those interviews, I was never contacted again. When I tried to inquire about the status of the dean's decision, she did not have the decency to return my calls. She hired an individual for the position who had not even finished the requirements for the doctoral degree, and therefore lacked the appropriate credentials for this position. This "professor" who terrorized both undergraduate and graduate students caused one graduate student to drop out of school, and finally left the school under controversial circumstances.

On a second occasion I interviewed for a similar teaching position at the Philadelphia School of Pharmacy and Science. During the day of the interview, I had to meet with 16 different professors and give a one hour seminar. The purpose of the seminar was to evaluate my teaching abilities. At the end of the seminar the chairman of the search committee shook my hand, smiled and said that I can teach. As we were walking down the hall to meet my next interviewer he said that so far everyone I met had liked me. During the seminar I summarized the work in cancer I had done to date and my future research plans which included a novel approach to the development of anti-cancer drugs. I made it clear to all of the faculty members I met with that the school already had the equipment I needed for my project. I also let them know that I was willing to teach any courses in chemistry they wanted, and suggested some new courses for graduate students. Several of the faculty members I met with on that day told me I had an "excellent reputation" at that school. In short, I was qualified, accommodating and responsive to their needs. I received a rejection letter about a week after the interview.

From all of these rejections, one of two conclusions can be drawn. First, that I am very difficult to work with. This is possible, but based on my history of involvement with team sports, high marks in job evaluations and strong letters of recommendation, I do not believe this is the case. The second conclusion is that none of the universities and associated teaching hospitals is interested in my work. Based on the similar experiences of other innovative scientists, I think I can provide an explanation for their decisions.

Academic Arrogance

The ego of a university research head can be reflected in at least two different responses to an aspiring, young cancer researcher. One response is to view the young scientist as a threat to his position on the staff. Suppose you have a novel approach to the development of anti-cancer drugs and need a place to work. Only a few places in the country are suitably equipped. If your plan is substantially different from the established researcher's, what do you think his response will be? From your perspective you wish to further the understanding of cancer. You may hope that if your idea was not as successful as desired, then at least it could inspire someone else to find the cure. From the established researcher's perspective, you have inadvertently told him that he is a failure in his field. He has spent the last 25 years or more of his life working on his project in a direction which is far astride the direction you want to pursue. All he needs is a little more time to iron out the details of his cure. He has been saying this for the last 15 years or more, and getting away with it. From his point of view, if you are right, then he must have always been wrong. Your enthusiasm to bring new ideas into the realm of cancer research will be interpreted as a threat to his tenure and leadership. The result is that he will not allow you to join his staff, especially if there is a good chance that you are right.

This is one of the reasons I can not go to a cancer research center and work on novel ideas for the cure of cancer. My ideas will not be in keeping with the thrust of the time worn plans of the regime. The fact that I might very well be successful works against me. In addition to the ego of the researcher being bruised, his grant funding will be placed in jeopardy. If someone else is right, then there is no reason to continue giving him money.

The second type of egotistical response results from the professor/researcher seeing himself as the center of the universe in his field. He and a small handful of others are the only ones who can truly understand the subject of his reputed expertise. When approached with your research plans, his response probably will be a question about why you can see the situation differently than his "distinguished" colleagues. Regardless of your answer and enthusiasm, he has already decided that you do not have the necessary credentials to carry out the work. Further, he will view your ideas as speculative at best. After all, he and his colleagues, each with over 20 years experience in cancer research, have not come up with the solution. How can you? Again, you may be completely right, but he will not take a chance in offending his colleagues. You will not be hired to teach at the university.

One professor I know once told me that no one will ever unravel the mechanism of cancer, because it is beyond him. The thinking here is typical: the "experts" can not do it, therefore no one can. They have certainly documented that they in fact can not unravel the mechanism of cancer or develop a cure, but that does not mean that no else can. They are limited in intelligience, but apparently not in ego.

CHAPTER 8

THE FEDERAL GOVERNMENT

It is easy to criticize the role that the federal government plays in cancer research. The two comments I hear most often are:

"The government is suppressing the cure for cancer; it should be spending more on cancer research."

"The government is conspiring with physicians to withhold the cure because physicians make more money treating cancer than if they were to cure the disease."

In my opinion and experience, the federal government is not withholding any cure for cancer and it is spending far too much money on cancer research. In fact, the government is cooperating with universities and private pharmaceutical companies to work on new treatment strategies. Our government allocated about 1.5 billion dollars to cancer research in 1991 alone. It also has three large agencies that are actively working on various aspects of cancer. These agencies are the National Institutes of Health (NIH), the National Cancer Institute (NCI) and the Centers for Disease Control (CDC). With all the tax money being spent on cancer research and the beehive of activity at government facilities it is fair to ask: What is wrong? Why can't they tell us the mechanism of cancer? Why haven't they developed any consistent cure for this disease? The answers to these questions are complex, but it can be boiled down to one simple answer. The wrong people are in charge of the research programs.

We can divide the federally guided personnel into two groups. Group 1 is comprised of the politicians such as the President, his Cabinet and the Congress. Group 2 is made up of researchers. The politicians in Group 1 are responsible for the allocation of funds to the numerous cancer research projects throughout our country. Beneficiaries include universities, government facilities and private companies. The politicians decide how much money is to go to every research project the government is interested in. These research projects include the mechanism of cancer, new treatments, and ways to prevent cancer. This huge expenditure of time and money strongly suggests that the politicians are actively waging war on cancer.

Politicians and Cancer Research

Let us assume for the moment that there is a conspiracy by our government to withhold the cure for cancer from us. It is intuitively obvious then that this cure must lie in a carefully guarded vault somewhere in this country. The solution is simple. All we have to do is find this secret hiding place, storm the facility, and obtain the documents for the cure. We could then manufacture the drug ourselves and dispense it to those who need it. We would charge the patients only the amount of money necessary to recover our expenses for making the drug. In this way we could break the cartel between the government and physicians.

However, no such document for the cure exists. Some of our top government officials have died from cancer or are currently suffering from this disease. If the cure existed, they would be the first to receive it. President Johnson and his former Vice-President Hubert H. Humphrey died of cancer. Senator Cranston is currently suffering from cancer. President Ronald Reagan had polyps removed from his colon and a small skin cancer tumor from his nose. The only real advantage that government officials have in the fight against cancer is that they receive earlier treatments for the disease than you probably would. The reason for this early diagnosis is that they are more closely monitored for medical problems than you are. Former Vice-President Humphrey even submitted to new cancer treatments as he neared the end of his life. He agreed to act as a guinea pig in order to further the understanding of the treatments for this disease. If politicians in high ranking positions die of cancer, then it is certain that there is no cure tucked away somewhere.

No amount of money can buy a cure for cancer. In the late 1970's the Shah of Iran came to this country to have his cancer treated. He was a multimillionaire who could afford the best medical care. He died of cancer. A well known and liked Hollywood personality, Michael Landon, recently died of inoperable cancer of the pancreas and liver. He is another millionaire who could afford the best treatment. The very rich and the very powerful die of cancer at about the same rate as the poor because there is no cure for cancer at this time.

Let us follow this government conspiracy idea further. Again we will assume that our government is withholding the cure for cancer. There is no doubt that our officials might be able to prevent the release of the cure to the general population of this country, but they have no control over what can be released in other countries. Almost every country in this world is funding some form of cancer research. For example, Sri Lanka has a very active cancer research program. The former Soviet Union spends huge sums of money on cancer research every year. Their cancer research programs rival ours. Yet they have made no more progress in this area than we have. It would have been a tremendous political advantage for the Russians to release the cure for cancer first. The same reasoning would apply to the Chinese, Japanese and all of the European countries. There would be no way for the United States to stop them from releasing the cure for cancer. No one in any country is bragging at this time.

I hope that I have reassured you that there is no government conspiracy to withhold the cure for cancer. Goverment officials die of this disease just as readily as we do. Rather than a conspiracy, it is quite clear that our government has constructed an elaborate system to overtly wage war on cancer. Our government funds research at universities, works closely with private companies and will even fund small businesses through its Small Business Innovation Research (SBIR) grant program in efforts to unravel the mechanism, to develop a cure or to prevent the formation of cancer. I can not imagine how our government can improve on its desire to defeat cancer. Nevertheless, something definitely is wrong with the system. The fault, though, is not with the politicians.

The Politics of Cancer Research

No one can purchase creative insights into the solution of a problem. Cancer certainly is a problem. The individuals who are working on this problem have well documented their failures. Volumes of publications in cancer related journals actually show how *unsuccessful* they have been. They have proven their inadequacy. All that money and time spent, and they still can not answer the basic question of what the mechanism of cancer is.

The solutions to the problems in cancer can only come from creative insight. Not everyone has this talent. No matter how much tax money we spend, this insight is clearly lacking in the researchers currently running the government funded programs. This result is not unexpected when you consider that the government employees who call themselves scientists are recruited from the same pool of physicians, molecular biologists and chemists already discussed. If you give more money to these researchers, it will be wasted. They will use the money to buy expensive new toys for their laboratories. A chunk of the money will go to higher salaries. They will take expensive paid vacations in the form of conventions in beautiful cities like San Francisco or New Orleans, insisting that they must attend to keep abreast of their field. Their alleged creative capabilities will not be increased by greater sums of money. The solution to unraveling the mechanism of cancer and to the development of cures is not in giving these failures more money to squander. The solution is to get rid of these egotists and to bring in fresh minds with new ideas.

The lack of real progress in cancer research is due not to the politicians who provide the money, but to these scientists who misuse it. The ego of these scientists blocks the advances in cancer research. The heads of the research projects invariably have their own "pet" theories. They will not even consider hiring anyone who does not share their view. Consequently, we see stale theories being rehashed in new forms. Most of their theories have no basis in fact. The situation is most often the opposite; the facts debase their theories.

To understand how and why this situation exists in government sponsored research, realize that the people who head these programs are most often physicains, not scientists. And, as explained earlier, the other major group of researchers, molecular biologists, do not understand the very discipline in which they claim expertise. What we have is truly a situation of the blind leading the blind. They can get away with their policies because until now there was no one in a position to challenge their authority. The heads employed by the various government agencies such as NIH, NCI and CDC are conning the world with their alleged knowledge and advancements in cancer. Some of these individuals have been involved in cancer research for over 20 years. They still can not tell us what the mechanism is or provide a cure for the disease. This is not an impressive track record. You should know in greater detail how these con artists have been spending your tax money over the years.

Physicians employed by the government typically surround themselves with scientists from whom they routinely steal ideas to claim as their own. My work experience at the Oak Ridge National Laboratories can illustrate my point.

Back in 1971, President Richard M. Nixon declared war on cancer. His administration recruited some of the top names in cancer research from throughout this country and the world. These scientists were assembled at the Oak Ridge National Laboratories in Tennessee. They had virtually unlimited resources to carry out their work. Part of this bill was paid by the Dow Chemical Company. The rest was picked up by the taxpayers. This facility was sold later to Martin-Marrietta. Physicians and Ph.D.'s were suppose to work on this project together in a joint venture. The level of funding and the human resources committed to this project indicate how serious our government is in pursuing a cure for cancer.

I do not know the exact figures for the operating budget at that time. I can describe for you the equipment purchased and the facilities renovated that were never used. Dozens of physicians and Ph.D.'s worked with scores of technicians and other support personnel such as secretaries and maintenance mechanics. I was told by one of the scientists that over 600 people worked at this facility at one time. You can imagine the size of the payroll alone. I found six storage rooms about the size of an average living room piled to the ceiling with crates of animal cages that were never opened. In the laboratories there were pieces of sophisticated scientific equipment just collecting dust. Some of the equipment had been unpacked but never used. There was one floor in the building that was specially renovated to house animals in a carefully controlled, sterile environment. The doors to the rooms had long rubber gloves in the front of them so that the animals could be handled without direct contact by human hands. I was told that the renovation of this floor cost about $5 million back in the 1970's. The cost would be substantially higher today. The floor was never used. Millions of tax dollars squandered just at this one government supported facility.

What kind of work did the researchers produce with so much of our money? Perhaps another way to phrase this question is to ask if we got a return for our investment. The simple answer is no. *No major breakthrough in cancer research occurred from work conducted at Oak Ridge.* Most of the articles I read on research done there were based upon a simple formula. Get an animal and a chemical. Give this chemical to the animal at the highest dose it can tolerate short of dying, and wait for about a year before it is butchered. At the end of one year, kill the animal and remove the organs for examination. If tumors are found the chemical is labelled a carcinogen. If no tumors are found the first time around, this chemical can be co-dosed with a known chemical carcinogen and labelled as a cancer-suspect agent. If only damage to an organ is found, then this result is simply reported. It is rare to find an article that reports no damage because the animals were given such high doses that some form of internal damage invariably occured.

Merely going through a list of chemicals to determine which ones are carcinogenic is not innovative research. Hundreds of articles were published by the scientists who worked at Oak Ridge, but no one could explain why a given chemical is carcinogenic or what the mechanism of cancer is. Millions of our tax dollars were spent on a futile project. Yet the self serving researchers brag of the accomplishments they made in cancer research. They tout their failures as though they acheived some important goal.

Eventually the cancer research at Oak Ridge was virtually terminated. Most of the Biology Division was shut down by an administrator from Martin-Marietta. He was no longer amused by the lack of results. From what I was told, he felt the same way that I did about the work conducted there. The waste of two decades and hundreds of millions of dollars with no important acheivements could no longer be tolerated. That administrator, although not a scientist, made the right decision. Apparently it was clear to him that simply dosing animals with the highest tolerated amount of chemical was not the way to unravel the mechanism of cancer. Nor could splicing DNA fragments into the genetic code of bacteria offer any hope for a cure. A newer and more creative approach had to be tried, but the scientists working there were not capable of innovative research. Oak Ridge is typical of the way government agencies did, and unfortunately still do, conduct ineffective cancer research.

The Federal Grant Game

Our government has a number of grant programs designed for small businesses that want to work on innovative ideas in cancer research. The problem with these programs is that the applicants must submit their research plans to the same failures who head the various government agencies. Those who review the grant proposals are in a position to take one of two actions. The first is to reject the proposal outright. They will do this especially when the ideas presented in the grant do not coincide with the fad theories of theirs and their colleagues. Perhaps you find it hard to believe that there are fads in scientific research comparable to the pet rock and mood rings. Fads in scientific research do indeed exist. Searching for oncogenes is one fad by which many researchers are padding their bank accounts.

The second path the reviewer can take is to reject the proposal despite recognizing that the ideas are based solidly on good data and scientific principles. Because the reviewer never could have developed these ideas, he rejects the proposal on some contrived premise, rewords the ideas presented in the proposal, and then lays claim to the ideas as his own. The applicant has no legal rights to challenge this theft by the reviewer, and the reviewer knows this. The reality of what happens when grant proposals are submitted is that bad ideas are rejected, and good ideas are rejected and then stolen. While the motivation of the legislative branch of the government to fund small businesses is noble, the implementation of this plan is quickly corrupted by the scientists who head cancer research agencies.

Theft of ideas is not the only despicable act committed by heads of government agencies. These researchers exert a tremendous amount of influence on deciding which articles can be published in their magazines, such

Cancer, Cancer Research and The Cell, as well as sneaking into scientific journals. This is a cleverly disguised form of censorship. In this capacity, they can control the flow of information to scientists by rejecting articles that challenge their theories, and accepting ones that support them. Even the egotists at universities are subjugated by these tyrants, although professors readily agree to this submission. The heads of government agencies exert the same dictatorial influence on the distribution of scientific information as on the projects funded.

When this theft of ideas and censorship occurs, you lose. You lose in two major ways. First, you end up paying for poor research with your tax dollars. Second, you pay with your life if you develop cancer. The thief can not successfully develop the stolen ideas. Only the person who submitted the grant application truly understands all aspects of the proposed work. It is this person alone who is capable of continuing the work by being mentally prepared to solve problems should unexpected nuiances crop up. The thief will get the recognition for the innovative ideas and will gain status in the community of cancer researchers. His power base will be extended while the rest of us are left to suffer and die from cancer. He does not care because his ego and status will be maintained. The individuals who head large cancer research facilities are egotistical miscreants who should be barred from the laboratory, not be put in charge of one. All of them have at least 20 years of experience in cancer research and what have they accomplished? They should not hold their heads high with pride. They should hang their heads in shame for being the incomparable failures they are.

Good Intentions Are Not Enough

The federal government, with good intentions, is doing all that it knows how to fight cancer. Legislators are funding projects to unravel the mechanism of cancer with the hope that this knowledge will lead to a cure. They also are funding projects directly working on curing cancer, and projects exploring ways to prevent cancer. The government is cooperating with private industry, universities, and their own facilities. Stagnation in cancer research that we face today despite this heroic and expensive government effort is due to the physicians, molecular biologists and chemists who hold high positions in federal agencies. They pursue their pet theories to the exclusion of more promising work. They exclude many talented scientists from the inner circle of cancer research for fear of being exposed for the frauds they are. Their strategy is to protect their self acclaimed exalted positions at the expense of our money and lives. This conspiracy of egos among the individuals in power is preventing progress toward curing cancer.

CHAPTER 9

PHARMACEUTICAL COMPANIES

Some of you believe that pharmaceutical companies have not produced a cure for cancer because they make more money selling drugs that treat the disease rather than cure it. The fact is that pharmaceutical companies do not make much money on cancer treatments. There is a limited market. Most patients die soon after they have been diagnosed, so there are few users of these drugs. For this reason, many pharmaceutical companies do not even bother to invest in the development of new anti-cancer treatments.

It would certainly be in the best financial interest of any company to come out with a drug that consistently cures cancer. That company could charge whatever price the market will bear. This price would be quite high as the drug AZT (zidovudine), currently used for AIDS, illustrates. The pharmaceutical company that markets this drug sells millions of dollars worth each year. This drug only treats, not cures, the disease, yet people willingly pay $800-$1,000 per treatment. Cancer patients would undoubtedly pay virtually any price for a drug that cures their disease. Let us examine how pharmaceutical companies operate to understand why they can't cure cancer.

High Stakes

Pharmaceutical companies are just like any other type of business. They try to market products that bring in money. Before a new drug can be put on the market, the company has to submit the compound to a series of tests required by the FDA. Guidelines have been drafted by the FDA to ensure the quality and the performance of the drug. These guidelines serve to protect us from fraudulent claims for products that may do more harm than good. With these tests to meet FDA requirements for approval, along with the expense of development, it can cost the pharmaceutical company $250-$500 million just to bring the drug to market. A major financial consideration for any company is the decision as to what kind of drug they want to develop.

On average, it costs a pharmaceutical company about $50,000 per chemical compound for the first round of tests to evaluate product safety and efficacy. If a compound is found to be unsafe at this first check point, then no further tests are performed and the compound is dropped from the list as a potential drug for human or animal use. Routinely, several thousand compounds are synthesized and tested. Even if only 100 compounds were tested initially, the company would spend $5 million just to complete the first round of testing. It is likely that none of these compounds would even pass this first test. To improve the odds, most often many more than 100 compounds are synthesized. Out of several hundred if not thousands, perhaps 10 compounds will pass the first test. Each of the subsequent rounds of testing becomes increasingly more complex and costly. Of these 10 compounds, the company will consider itself fortunate should one eventually get FDA approval for marketing.

Some of the tests required by the FDA are bypassed with anti-cancer drugs. The thinking is simple but appropriate. If nothing is done to try and prevent the growth of the cancer, then the patient will definitely die from the disease. It then is justifiable to try any new drug with the hope of having some effect against the cancer. If the new compound was tried on laboratory animals with some measure of success at inhibiting tumor growth, then it is tried on volunteer human subjects. The patients are told beforehand that the treatment is experimental. The drug is used with full patient consent. If this new drug prolongs the life of the patient, then it can receive FDA approval for general prescription use against cancer. The FDA, at this point, is not as concerned with negative side effects, like birth defects or hair loss as it is with other drugs because the patient probably will not live long enough to experience such side effects.

There are several pharmaceutical companies that have anti-cancer drug projects. The motivation for these projects is financial. As of this date, there are no compounds which consistently and reproducibly cure any type of cancer. Consequently, the company that comes out with such a drug will corner the market. That company can charge any price it chooses. These companies are therefore pouring tens of millions of dollars into their anti-cancer research gamble. They have hired who they feel are the best researchers in the world. Yet, even with such a huge investment of resources, none of their people can develop a cure for cancer. Ego is once again the root of their well documented failures.

The Guru System

Pharmaceutical companies, like other businesses, preach the "team concept" to their employees in an effort for all to cooperate in performing assigned tasks. It is a nice theory, but it quickly breaks down in their research environment. Team concept becomes translated into supporting a self proclaimed "guru" who sits at the head of a research project or group of scientists within the company. This functional idiot sees to it that only those individuals are hired who support his pet theories. Even at the Ph.D. level, he expects everyone to make him look good to company executives. This guru may have been at the cutting edge of research at one time, but new ideas have passed him by. He will not, however, relinquish control of his group to any new researcher with a fresh approach to the project for fear of losing his position. As a result, the company keeps rehashing old ideas that have never produced a successful anti-cancer drug. Just the simple act of suggesting a new idea to his group can set off a job threatening defensive response by the guru. It is not in the best interest of a young scientist to even hint at a challenge to his authority. I do know of incidences whereby a guru got rid of the very scientists who helped to put him in his position because he now viewed his supporters as threats who could expose him as a fraud. There are extremely egotistical individuals in charge of research departments in pharmaceutical companies. These malevolent personalities hide behind company policies to present an official front of cooperation.

Pharmaceutical companies are perhaps the most schizoid of all businesses. They go from feast to famine in huge swings. While they own the patent rights to a major drug, they make billions of dollars a year. When the life of the patent runs out, there is a great panic. It is during this time when the gurus often are discovered for the frauds they are. It becomes evident to the executives who actually run the company that the guru only wants to pursue his own ideas. Ordinarily there is nothing wrong with this approach as long as his ideas lead to new products. Because the guru surrounded himself with sycophants, however, no new ideas can be presented. Consequently, the guru often is let go and there is a major restructuring within the company. Other scientists are also released and fresh talent is brought in. This reaction is understandable and may be appropriate, but the companies recruit the new researchers from the same pool of personnel discussed in previous chapters. These individuals come from the same self-serving physicians, molecular biologists and chemists destined for failure in cancer research. The top eschelon of officers running a pharmaceutical company may want to improve its research program, but they can not because the talent is just not available.

When the major reorganisation within the company is finished, the new regime may not continue to support previous projects. Some companies may even drop cancer research. Another course of action may be to shift the focus of cancer research from anti-cancer drugs to the development of other agents reported to have alleged anti-cancer properties, such as interferon, or to alternative delivery systems for existing drugs.

Ideas Non Grata

You might wonder how decisions to set or change the thrust of research projects are made. It would be nice to say that learned scientists carefully review the scientific literature on a given proposal to determine the feasibility of the project. Or that group leaders evaluate the qualifications of personnel in order to put the right person on the job. Or that all of the different departments associated with the project worked in harmony to achieve the desired goal of putting out an anti-cancer drug. This scenario does not describe the real situation.

A tremendous amount of petty interoffice politics comes into play when choosing the projects to be funded.

The most politically powerful person, not the most intellectually competent, usually sets the course for the research to take. A pharmaceutical company may even bring in a Nobel Prize winner to head the cancer research project. But despite impressive credentials that include a long list of publications, he can not explain what the mechanism of cancer is or why his earlier work has not led to curing this disease. The answer to these questions is that he has no new ideas in cancer research. He is resting on the laurels of his past. His great promise as a cancer researcher is only an illusion, which the pharmaceutical company naively embraces.

I know from personal experience that pharmaceutical companies are not interested in new strategies for anti-cancer drug design unless some group leader in the company can take credit for the idea. Some time after I felt confident that I had unraveled the mechanism of cancer, I hired a patent attorney to contact three pharmaceutical companies. I thought that a letter from a patent attorney to the presidents of those companies would have more impact than sending a letter on my own behalf. The Sterling Pharmaceutical Division of Kodak Chemical Company never replied to my proposal. DuPont and SmithKline-Beckman both engaged me in legal dialogue through my attorney.

Here is the deal both companies offered. I was suppose to send them my exact strategies for the design and development of anti-cancer drugs, including my theory on the rationale for those designs. They would have their experts evaluate the feasiblity of my designs. If the company rejected my proposal, then it would not work on the development of these potential drugs for a period of 5 years, and here is the kicker, *unless* their scientists were already pursuing research in the area which I proposed. At no time did either company actually indicate that it would be interested in hiring me to pursue my novel approach. After I sent each company a 10 page review explaining why current cancer treatments must necessarily fail (which is included as two expanded chapters in this book), I requested a meeting with an appropriate company representative to discuss the possibility of joining that company to continue my cancer research with them. Both companies stopped the legal dialogue at that point.

The actions taken by these companies can be interpreted a number of ways. Based upon by experience with large chemical companies and with other cancer researchers who tried to steal my ideas, I concluded that these companies simply wanted something for nothing. They thought they saw a sucker coming, but I had anticipated their tricks. Obviously, the "scientists" who would have reviewed my proposal thought I was as stupid as they are. I never released all of the information to any member of any pharmaceutical company.

Let me explain why I feel that these companies tried to steal my ideas. Each had, and still have, an active research program to develop new anti-cancer drugs. This is one of the reasons I selected these companies. The program in each company was, and still is, failing miserably. An obvious indication of this statement is that none of the companies have announced a cure for cancer to date. In the case of SmithKline, the patent for their number one selling drug, Tagamet for ulcer treatment, had run out. Other pharmaceutical companies were already marketing competitive anti-ulcer drugs. SmithKline was in an almost desperate situation to come up with a new patented drug to pick up the slack in company sales. And what could be a bigger winner than a cancer cure? After I submitted my proposal, SmithKline Beckman reorganized to become SmithKline Beecham.

This corporate shuffling is a typical reaction to a company's loss in prosperity, aggravated by failure to market the next "big new drug".

Three years after submitting my proposal to DuPont, I went for a job interview with the group of medicinal chemists responsible for the development of anti-cancer drugs. This company's program is still failing miserably. They have no new drugs to offer that provide any advantage over the existing drugs currently being used on cancer patients even though an attempt to develop one is still being conducted.

The plan at these companies was, in my opinion, to obtain my strategy for a novel approach to anti-cancer drug development. Their experts would review the proposal and, of course, they would reject it for any number of reasons. Meanwhile, these same experts would start to explore my ideas without my knowledge or permission. They would try to skirt my legal claims to their work by rewording the concepts I presented, assuming they had the intelligience to do so. It would have been difficult for them to alter my ideas without being obvious. At the very least, they would have involved me in a lawsuit which they would plan on winning because their legal resources are much greater than mine. The other alternative was to just sit tight on the ideas presented in the proposal until the 5 years had expired. The reality that millions of people would suffer and die from cancer during this time period was of no concern for them. At that point, they would be free to openly develop my ideas without me because I would have signed away my rights in the confidentiality agreement their attorneys had sent.

If either company I contacted, or any other company, had decided to work with me on the novel and rational design of anti-cancer drugs, by now there would have been an experimental drug on the market effective against the lung cancers seen in tobacco users. In my estimation, millions of patients worldwide could have been cured by now. Instead, due to the egos of the cancer researchers in the pharmaceutical industry and to the greed of those companies, millions of people throughout the world are needlessly suffering and dying every day. Lung cancer from tobacco products is the number one cause of cancer death in the United States. When lung cancer is diagnosed, it most often is inoperable and the patient dies a relatively quick but painful death. Everyone loses due to the petty vanity and greed by employees of pharmaceutcal companies.

Costly Futility

Researchers employed by these companies have only a little more understanding of cancer than you do. They can not unravel the mechanism of cancer nor can they develop any effective cures for this disease. You are paying for their ineffective cancer research. The way pharmaceutical companies squander money is almost incomprehensible to the average businessperson. These companies have to charge higher prices on products not associated with cancer in order to fund their cancer research programs. You pay for this research when you buy an antacid product or make-up, for example. More importantly, you are again paying with your lives because the gurus employed by these companies are not receptive to new ideas unless they can lay claim to those that might work.

You now have an understanding of why pharmaceutical companies have not been able to develop cures for cancer. These companies would like very much to market a cure if they could. They have even more to gain financially from an anti-cancer drug than physicians and hospitals have to gain from merely treating cancer.

Pharmaceutical companies are another example of the wrong people working on the problem of cancer.

CHAPTER 10

DUCK HUNTING

Many of you believe that cancer research has to be very expensive. You gladly give money to organizations like the American Cancer Society in the hope they will unravel the mechanism of cancer, find a cure for the disease, or discover ways to prevent it. These organizations bombard you with high pressure ad campaigns to give more money because the answers to this disease are "just around the corner". If they only had more money, then all of the questions would be answered, in time. Notice that members of these organizations never specify when that time will come. This leaves them an easy loophole should you ever ask. Definitive sounding allusions to future gains is one of the strategies typically used by con artists. This is one of the reasons I refer to cancer researchers as con artists, because they routinely use this tactic. Often you also will hear such slogans as "we offer hope because we ask questions". I am not interested in anyone asking questions. The substance of the questions is obvious: cure my disease. Let me help you understand how cancer researchers spend billions of your tax dollars without giving you any answers to the questions they claim they are asking.

I like to make the comparison between cancer research and duck hunting. Let's assume that killing a duck is in the best interest of national security. For you animal rights activists who might be offended by this action, I agree that this is an unpleasant task. However, just as you have no say in the matter of our government spending money on a duck hunt, so too no cancer patient had any say in the development of cancer. No woman woke up one morning and decided to develop that breast cancer she had been dreaming of for years. A man does not plan for two decades to become diagnosed as having lung cancer. This is not the motivation for his smoking habit. In an analogous manner, we have no choice but to financially support our government's quest to kill a duck.

Once we agree that killing a duck is in the national interest, it is logical to ask what the best way is to perform this task. Whatever strategy we adopt can be applied to both unraveling the mechanism of cancer or to the development of its cure. I can envision at least two approaches to this assignment. One is similar to the way cancer research is conducted up to the present time, which I call the shotgun method, and the other is how I can do it. The first way will cost you billions of dollars with no guarantee of success. My way will cost you a few million dollars with a near certainty of success. Let me describe both approaches to you, and then you can make up your mind as to which approach is better.

The Shotgun Approach

I call their approach to research the shotgun method due to the similarities to the design and use of a shotgun shell. A shotgun shell appears to be a formidable weapon because it is about three-fourths of an inch in diameter and about three inches long. The shell can contain several hundred pellets. After the gun is fired, the pellets spread out such that the farther they travel, the broader the scatter pattern. The shotgun is designed for the close range killing of small animals. While it can create a tremendous amount of damage at close range, it is ineffective at distances greater than 100 yards. This gun makes a lot of noise and can do damage, but it is shortranged. Therefore, the utility of a shotgun as a weapon is limited in scope.

The same senario is true for cancer research. The projects and ideas presented cause a lot of commotion, but their utility is short-sighted and ineffective for the tasks at hand, which are to unravel the mechanism of cancer in order to develop cures for the disease.

The main rationale of almost all cancer researchers is analogous to the shotgun blast in that they want to fire ideas in all directions in the hopes that at least one will work. Cancer researchers do not have as well thought out a plan for conducting their projects as they like to present to the public. They come up with hair-brained schemes which have no possibility of working. These failures could be avoided if only someone took the time to logically critique their work. Instead, the guru of a project bullies his subordinates into following his master plan.

One of the ideas that has gained millions of dollars of support is the notion that interferons can cure cancer. There really is no data to support this notion. Because the failures are frequent and consistent, the researchers try to cover up by claiming that they need more money to isolate purer and thereby more effective forms of this toxin. To date, no one has been cured of cancer using any form of interferon. The cancer patients merely go into a state of remission for varying lengths of time depending on the dosing regimen and individual patient response.

The researchers redefine a "successful cure" by the average extension of life for patients treated with interferon.

Duck Hunting Committee

I want to tell you how a typical cancer researcher would attempt to hunt a duck. You then can understand how easily and rapidly cancer research can become expensive. Let us assume that the federal government has made the decision to hunt a duck. Legislators will try to find the best candidate for the position of chief duck hunter. Choosing this person will probably be based on the number of scientific articles that person has published in biologically related journals. The reason for this choice is simple. Ducks are animals, therefore they fall into the general category of biology. A biologist should have the best understanding of organisms, including a duck. The legislators will therefore choose a biologist to head the duck hunting project. Of course another major reason for selecting that particular biologist is that he knows someone influential in the government. We are not suppose to acknowledge this fact of life, but denying it will not change the situation.

If the chief biologist is like the heads of cancer research programs in this country, he probably will not know what a duck is or what it looks like. Consequently, he will have to hire someone else to describe the animal to him. Through the "good old boy" system of hiring individuals in high positions, he will get another biologist who has heard something about the way ducks look, but is not sure. Then these two biologists will hire another biologist who may actually know what a duck looks like. Very quickly a committee is formed. The reason why neither of the other two biologists were hired to direct the program is that neither had the political connections of the first. The salary for the chief biologist could be $150,000 per year. The salary for the second biologist could be $100,000 and the third biologist could receive a salary of $75,000. The initial cost for the duck hunting project is $325,000 per year just for salaries. There is no guarantee at this time into the project that they will succeed in even finding a duck.

After the committee is formed, one of the first items on the agenda is to agree on what a duck looks like. This will take some time to research, perhaps months. During the course of this research the biologists might even discover that ducks can fly.

The next item on the agenda is to agree on the location to hunt the duck. They choose downtown Manhattan because they want to live and work in New York. It is convenient to major universities and libraries and promises an entertaining social life. Pretending to want to keep the research costs at a minimum, the biologists agree to use a shotgun to kill the duck. It may not be the most efficient weapon, but the biologists stubbornly do not want to consider any alternatives.

The thrust of the project seems straightforward at this stage. All they have to do is fire the shotgun in as many directions as possible and as often as possible. The rationale for this strategy is simple. None of the committee members can predict when a duck will fly into Manhatttan or the direction of flight. By firing the shotgun as often as possible, they feel certain that a duck can not get past them. But as fate would have it, no duck is killed using this tactic for a month. Meanwhile, the biologists have gone through 50,000 rounds of shells at a cost of 50 cents each for a total price to date of $25,000.

Big Guns, Little Results

It becomes clear to the biologists why they were unsuccessful in killing a duck. They need more money for the project so that they can hire more personnel and buy more shotguns and ammunition. Members of Congress accept their argument and allocate more money to the duck hunting project. More staff is hired and ammunition bought. The shooting continues with more shotguns. More area of the sky is covered. Another month goes by without a duck being killed. One of the biologists overhears a deer hunter talking about the buck he killed at a distance of 500 yards. The biologist has a brilliant idea. The reason for the group's failure so far is that the shotgun may be too short-ranged a weapon to hunt a duck. Ducks may be flying out of the reach of the shotguns.

The biologist discusses this idea with the other biologists who heartily agree. They have a problem though in that they do not have sufficient funds to buy longer range weapons. Again the biologists write a grant proposal to ask Congress for more money to work on this new plan of attack. Someone in Congress points out that the military has long range weapons that can be used. Some of these weapons, however, such as tanks and missile launchers, need specially trained personnel to fire them. A battalion is assigned to the biologists to help them hunt the duck in Manhattan. You can imagine how the price tag for this duck hunt sharply rises when the military is called in. One missile can cost as much as $100,000.

The scientists and an Army battalion set up camp in Central Park. They fire shotguns in every direction for a short range assault on the duck, rifles for a range comparable to the deer hunter, tanks for long range, and missiles for the maximum distance. This all-out assault lasts for a month. One month of constant firing of these weapons costs tens of millions of dollars. Still no duck is killed.

The next phase of the operation calls for ballistics experts to enter the project. These individuals will be responsible for the design of more accurate long range missiles. They will develop new computer equipment and software based on the information supplied to them by the biologists. Development of new computer equipment will take years to accomplish and cost hundreds of millions of dollars. At the end of this time there is still no guarantee that the duck will be killed because the design of the whole project was not based upon any scientific principles or established theories. The guiding principle of the biologists was based solely on the fact that ducks fly. This was far too simplistic an approach to take.

Let's say that after one billion dollars are spent, the biologists and Army finally kill a duck. Bullets, bombs and missiles were fired in every conceivable direction for some period of time. Finally a duck was brought down at great expense.

Overkill

Besides the cost involved, there are several aspects of conducting a research project with this approach that I find troublesome. After all the time and money spent to kill the duck, no new guiding principles or theories were developed that would aid in bringing down the cost of future duck hunts. The only strategy we still have is to fire as often as possible in as many directions as possible until a duck is killed. Maybe as a result of the duck hunt some significant technological advances have been developed that can have applications in other fields. But these advances provide no consistent, inexpensive and efficient technique for killing only ducks.

There is another problem I have with this approach to duck hunting. What do you suppose happens to the bullets and missiles which missed their target? Those projectiles land somewhere with the potential to cause considerable damage to buildings and to people. This is analogous to the grandiose claims of cancer researchers and their "promising new treatments". Huge sums have been invested in those con artists to develop these treatments. Some researchers have come up with new imaging techniques. Others have displayed lasers which they claim can zap the tumor away very precisely with minimal side effects. Still other researchers have presented very poisonous chemicals arbitrarily labelled as anti-cancer drugs. None of these gadgets and poisons can produce the claimed results. Their lasers, radiation and poisons kill every life form they come into contact with just like the indiscriminate use of bullets and missiles to kill a duck. None of the agents currently used to treat cancer are selectively toxic to the tumor cells, just as no single type of bullet or missile can selectively seek out ducks. Also, just as the biologists on the duck hunting project have no real understanding of how they actually brought the duck down, these so-called researchers have no real understanding of how most of the agents they are selling to the public as anti-cancer treatments kill life forms. Their "bullets and missiles" sham treatments also must land, and when they do they cause mass destruction to all types of human tissue. The shotgun approach to cancer research used by all cancer scientists is doomed to failure.

Some of you may be asking why such a stubborn, narrowminded attitude exists toward new ideas in cancer? I attended a lecture once given by the head of the cancer research program in one of the largest pharmaceutical companies in the world. He was asked this same question by a member of the audience. He replied simply that the cancer researchers do not know of any other way to treat the disease; therefore, they continue to vary the structures of current drugs and other agents.

Inexpensive, Effective Duck Hunting

I can envision a different approach to the problem of hunting a duck. There are at least three pieces of useful information. First, a sketch of a duck can be obtained easily and inexpensively. Second, a duck can fly. And third, one is justified in looking for a duck in Manhattan because it was recorded that ducks did fly through Manhattan at one time. But Manhattan may not be the best place to look. Using this information, I can kill a duck for you for about $150,000. Also, I can get this duck in less than a year. When I am through with this project, all of us will have a good understanding of how to kill ducks routinely should the need arise.

I justify the operating budget as follows. I want a salary of $50,000 per year. This salary is cheap for a highly educated and experienced investigator. I need a four-wheel drive vehicle to conduct my investigation. Such a vehicle costs $20,000. In anticipation of camping in the wilderness to find a duck, I will set aside $20,000 for camping gear and special pieces of equipment such as a video cassette recorder. The rest of the money will go for travel expenses including gas for the vehicle, food, lodging, etc. The money does not have to be spent all at once. Rather, I think it would be a better policy to spend this money as it becomes necessary. For $150,000 the government will have its duck.

To aid my search for the duck, I can make a few assumptions, that must be tested, about its most likely habitat. I can assume that a duck is a wild animal. I can make this assumption because I have seen very few pet ducks, especially in Manhattan. Personal experience has shown that this assumption is most likely correct. I could be wrong. Ducks could fly through Manhattan at night, which would make spotting them more difficult. Perhaps there may not be any ducks in Manhattan, but they may reside in other major cities. I could investigate this possibility by at least two methods. The first way would be to drive to each city I was interested in to ask local authorities if they have seen any ducks. This strategy would add a large expense to the operating budget. If I were like most cancer researchers, I would not care because the taxpayer would pay for the trips. I could write them off as necessary travel expenses and I could claim that I had to attend conventions in those cities. I could be pampered by staying at the most expensive hotel in each city and eat at the finest restaurants. Remember, I am on a vital mission for the government.

The other method I could use is to simply call the appropriate authorities in the cities I was interested in for the information I needed. This would save time and money while providing the same amount of information. It is not as exciting as the other method, but it is substantially cheaper and equally effective. It would take me a week to obtain this data. I would contact the cities from the various geographical regions of the country and the world. Once I got this data, it would be up to me as an experienced researcher to properly analyze it. The authorities in each city would report that they either saw or did not see ducks flying overhead. After enough data was collected, a pattern would emerge. Some geographical region might have a large incidence of flying ducks. If this became evident, then it would be an easy matter to visit that city or region. All I would have to do then is observe the flight pattern of the ducks for a period of time to learn their habits and to determine the best weapon to use.

In reality, we know that ducks do not fly through cities as a general rule. Probably I will have to go to a different type of habitat to find a duck. One point I am trying to make here is that I will not allow myself to be so linearly focused on one course of action that I will miss opportunities present in another.

Cities are one end of a spectrum of human populations. Cities are the most populated places on earth. Let's say I found no flying ducks in cities. It would therefore seem to me that an inverse relationship between human populations and duck populations exists. The more humans there are, the fewer ducks there seem to be. It could be possible, then, that the fewer humans there are, the more ducks there might be. I would then go to the other end of the population density, the wilderness. At this point, I could not be certain that I would find ducks there. I would want to explore both extremes first and then begin to narrow my choices on the human population spectrum.

Because there are no paved highways in the wilderness, I need a vehicle that can cross the rugged terrain. That is the reason for selecting a four-wheel drive vehicle. There are no buildings in the wilderness, so I need the camping gear and supplies for food and shelter.

The wilderness has different types of terrain. I would spend time in each until I either spotted a flying duck or I was convinced that no ducks live in the wilderness. If I did not see any ducks, then I would move to another type of habitat. This habitat would exist in the middle between a city and the wilderness such as a farm community. Gradually the possible places to find a duck would be narrowed.

After I routinely spotted flying ducks in the appropriate habitat, I would observe and photograph them for a period of time with the video cassette recorder. After observing ducks for some period of time, I would know their patterns of behavior. This information would enable me to choose the best time and weapon to kill the flying duck. There are a number of advantages to my methodology for duck hunting. First, the cost is very much less than the shotgun approach. Second, it is much more efficient and effective in that I will have arrived at the solution to the problem quicker than the others. Third, when I fire at the duck, I can target only that animal. No other animal species will be killed. Only the duck I target will be killed. Buildings and human lives will not be in danger.

There are other important advantages using my approach. By observing ducks in their natural habitat, I would learn that there are other species of birds that fly and swim like a duck, such as geese and swans, but that are different from ducks. It would be obvious to me that the same method and weapon used for the duck hunt might not be applicable to other species of water fowl. After observing the different types of water fowl in their natural environment, I could develop a theory on their behavioral characteristics that would enable anyone else to more efficiently hunt water fowl in general and not just a duck.

An analogous situation exists in cancer. Not all types of cancer are the same. There are tremendous differences in the biochemistry and pharmacology among the different types of cancers. We must speak of cancer in terms of the particular *cell type* which characterizes the tumor. For each cancer cell type, a different anti-cancer drug must be developed.

CHAPTER 11

BASIC RESEARCH

Scientific research can be divided into two broad categories: basic and applied. Basic research is a necessary attempt at answering fundamental questions about the subject investigated. Experiments are conducted to collect data for analysis. From this analysis principles are derived and theories developed. Applied research utilizes these theories to develop products. Most people have done some form of scientific research. Many of you have bought a car. Before you purchased that car, you asked yourselves a number of questions. You looked at your financial situation and your personal needs before you went for a test drive. You compared features and the price of each make of car you examined. This is basic research. When you decided on a make of car, you sought ways to finance your purchase. Getting money to buy your car can be thought of as applied research because you used the data you gained from looking at your personal needs and your financial situation to determine how much money you had to borrow. Scientists perform the same process, except they use sophisticated instruments and principles to examine more complex subjects.

The Basics

Fundamental questions had to be answered early in cancer research. Data had to be collected to discover any patterns that emerged concerning possible causes of cancer. Fifty or more years ago when scientists did not have the sophisticated instruments and theories available today to assist their fact gathering experiments, a shotgun approach to cancer research was justified. Back then very little was known about any aspect of this disease. Physicians had very little understanding of what caused cancer. Scientists knew even less about cancer than the physicians. It is not surprising that no cure was quickly forthcoming.

Crude sociological experiments to delineate factors that contribute to the formation of cancer were necessary back then. Physicians tried to find a link between cancer and places where people worked and lived, their eating habits, and the history of cancer in the family. With time, some patterns did emerge. Eventually scientists were able to demonstrate that some chemicals to which humans were exposed formed tumors in laboratory animals. Some physicians concluded that viruses caused cancer. Of course, now that molecular biologists have the dominant voice in cancer research, they believe that cancer is caused by genes.

In the early days of cancer research, examining any and every idea in an effort to understand the mechanism of the cause of cancer was certainly justified. This type of basic research was necessary because the alternative was to do nothing about this disease. What is not defensible is how basic cancer research has been conducted in the last twenty years. Virtually no new information about the disease has been obtained. No new fundamental principles have been established.

Nearly everyone involved in cancer research today would, of course, strongly argue these points. They point out all of the "advances" they have made, especially in the last 5 years. Physicians and molecular biologists brag about the strides they have made in understanding cancer by displaying the hoax they call oncogenes. Chemists, especially those who call themselves medicinal chemists, boast of the new "treatments" they have developed. Let us take a close look into the realities of "progress" in basic cancer research.

A random selection of any of a number of articles published in the many journals about cancer research demonstrate the poor quality of the current work. Rather than mentioning names to provide examples, identities will be withheld to protect the guilty and to avoid litigation. Instead, we can consider the types of experiments that are being reported by the various technical disciplines involved. These are the experiments conducted primarily by physicians, molecular biologists and chemists. Remember throughout this discussion that you are paying for these con artists to play in the laboratory. Remember also, these are the very individuals upon whom our government relies to provide the legislators with the information used to set policies.

MD's Method

Physicians conduct two general types of experiments. In one, they love to link some agent to cancer. They are the bright boys and girls who reported correlations between too much sex and cervical cancer and between too little sex and breast cancer. How a woman can determine just the right amount of sex to have in order to avoid both cervical and breast cancer is presumably still under study. I discussed in a previous chapter that recently one physician somehow linked high voltage power lines to leukemia and brain cancer in children. Again, I do not know how high voltage wires can distinquish between children and adults. Perhaps after the age of 18 people become immune to electricity. Perhaps power lines are afraid to mess around with adults. I am sure this physician can concoct some story to justify his research. If you buy this though, please let me know so that I can sell you a few bridges.

The second type of experiment popular with physicians is to grab a chemical and ram it into laboratory animals at the highest concentration possible before the animal dies from an overdose of the chemical itself. The physician then reports one of three possible results: 1. The chemical caused cancer; 2. The chemical did not cause cancer; or 3. The chemical did not cause cancer, but tissue damage occurred. It is unimportant to the physicians that a large number of animals may die from the chemical long before any tumors form. This brand of science takes about as much intelligience to perform as is required to fall off a log.

Physicians can sometimes be crafty. Suppose no tumors or tissue damage is observed after the animals are dosed by the chemical. How does the physician respond to this undesired result when he wants to label a chemical a carcinogen? The physician reponds by either falsifying data to show that tumors were found, or he co-doses the animals with a known chemical carcinogen. With the latter option, he can then report the chemical as a suspect carcinogen or a co-carcinogen. Who is going to challenge his results? Another physician might make some sort of a fuss, but he will be challenging his own kind. This type of protest is not likely to happen.

There is someone, or actually something called a corporation, that might challenge the report by the physician. That challenge is in the form of a series of experiments financed by the company that manufactures the chemical in order to refute the findings of the physician. The company and the physician enter into a scientific dialogue, in the form of published articles, over the chemical in question. This dialogue can last for years with no resolution. Guess who pays for this nonsense? You do, and twice. Once is with the tax dollars that funded the physician's research grant. The second time by paying higher prices for the products manufactured by the company so that it can support the experiments to refute the physician, and recover some of the income lost from the sale of the product until the issue is settled.

When any product gets linked to cancer, it is virtually the kiss of death. The FDA will pull it off the shelves immediately. Once a chemical is linked to cancer, it is extremely difficult for the manufacturer to reintroduce it to the market. This happened with monosodium glutamate (MSG). This is a naturally occurring chemical extracted from soybeans. The Chinese have been using this product for centuries to enhance the flavor of foods. Despite the best attempts by some cancer researchers to label MSG a carcinogen, the FDA could not find any real supporting evidence.

If those researchers knew any science, they would have concluded that MSG can not possibly cause cancer. Let us analyze the composition of MSG. Mono means one. Sodium is one of two elements that comprise ordinary table salt. We all eat table salt everyday without any risk of developing cancer. Glutamate comes from glutamine, one of the 21 essential amino acids. It is illogical to believe that a necessary nutrient for our bodies is a carcinogen. The only difference between glutamate and glutamine is that a hydrogen atom was removed from glutamine and replaced with one (mono) atom of sodium. You can consider MSG to be a chemical derivative of table salt. All of you who cook know that salt is used to enhance flavors, as is MSG. Because no one who conducted the experiments to link MSG to cancer knew any chemistry, or ignored what they knew, we were forced to spend millions of tax dollars on this bogus research.

Animal Sacrifice

Physicians have used a large variety of animals to conduct their experiments. There is a textbook that lists a number of chemical carcinogens and the different species of animals in which those chemicals were tested. In addition, they have tried many methods to ram numerous chemicals into animals, called routes of administration. These methods include injections into the gut, under the skin, down the throat, inhalation, and skin patches. Can you imagine how difficult it would be to dose a guppy? There was one researcher at Oak Ridge who worked with frogs. I inherited his laboratory space after he left. I do not know what type of experiments he conducted, but I can tell you he spent a small fortune on glass bowls of various sizes. There were hundreds of these bowls in the laboratory. If any of you need glass bowls, perhaps you can work out a deal with Martin-Marietta, the owner of the building.

I understand and agree with the need to determine which chemcials are carcinogenic. It is necessary to screen virtually all products for potential carcinogenic risks. However, merely screening chemicals for carcinogen potential is not creative research. It offers no possibility of unraveling the mechanism of cancer, especially the way the experiments are conducted. Ramming a bunch of chemicals into animals for a period of time and then watching those animals suffer and die is a waste of time, money, and effort.

This type of work can not add to the understanding of the mechanism of cancer, or provide promising leads for any cures. Using one chemical carcinogen to dose many species of animals does not give us an explanation of why that chemical is a carcinogen. Using several different chemical carcinogens on one or many species of animals also can not provide information on why those chemicals are carcinogenic. The only information that is provided is that the chemicals are carcinogenic or that they are not. It would seem more productive to take one chemical carcinogen and focus on one animal species to ascertain exactly why that chemical is carcinogenic. There is no need to butcher millions of animals for useless cancer research. When cancer researchers do try to focus on one chemical carcinogen and one species of animal, they also are severely criticized by their peers for the limited focus of their work.

In reality, no animals need to be killed anymore for the sake of basic cancer research. There already is available in the literature enough information for an intelligient individual to assemble all the pieces of the puzzle in order to unravel the mechanism of cancer. I already have, as you will read in a later chapter. For those working on basic research, tissue culture most often can substitute for animals. Tissue culture is the growing of live cells in a nutrient broth. I have conducted a number of experiments using human lung cancer cell lines growing in culture. Tissue culture, in the right hands, is actually more informative than working with animals. Carefully defined biochemical experiments can be conducted with tissue culture such that the exact enzyme responsible for the transformation of a chemical carcinogen can be determined. Tissue culture is also easier to work with, cheaper, more reliable, and produces results more rapidly than whole animals. When I worked for Bionetics

Research, Inc., my supervisor, a chemist, did not allow me to use tissue culture. I was forced to choose between having rats killed or unemployment.

I do not find killing animals pleasant. Scientists kill animals by one of three methods: 1. decapitation by a guillotine, 2. asphixiation with carbon dioxide, or 3. death by lethal injection, usually with a barbiturate. These killing techniques are considered to be humane because they are rapid and the animal allegedly does not suffer. When I had to kill Syrian golden hamsters by carbon dioxide asphixiation, a peculiar thing sometimes occurred. For this technique, the animal is placed into a large glass jar. Carbon dioxide is then allowed into the jar until the animal dies. Syrian golden hamsters have a habit of resuscitating soon after they come back into regular air. There were several occasions when I had the chest of the hamster open and the heart started to beat again. So much for the so-called humane methods.

There is no humane way to recklessly butcher animals. The carcass is thrown away, although we are not suppose to say this; we are supposed to say that the carcasses are disposed of properly. Disposing of the carcass properly means that it is placed into a plastic bag and thrown into an appropriately labelled trash can, which is removed by a specially licensed trash removing firm. Of course, if the animals were dosed with radioactively tagged chemical carcinogens, then the carcasses go into a different trash can for landfill dumping.

MB's Phantom Genes

In most cancer research laboratories today, physicians do not work alone. Molecular biologists and chemists are right by their side. Molecular biologists like to take the sum total of all of creation and reduce it down to some gene they allegedly found. Not too long ago, molecular biologists claimed that children are addicted to watching television due to a gene. Can you imagine that humankind's efforts at survival and the building of civilization were for the sole purpose of having our children become addicted to watching television? Humans struggled with the environment to survive. We learned how to plant crops and tame animals. Mighty civilizations came and went. All of this adaptation by humans just so that we could engineer into our genetic pool the gene for television watching addiction by children. This is typical of the absurd research and claims made by molecular biologists today. Molecular biologists are considered, nevertheless, the sacred cows of medical research. That they command a tremendous amount of power is the reason for the perpetuation of their useless theories and experiments.

Molecular biologists claim that cancer is caused by a set of genes called oncogenes, which code for the formation of cancer. According to their oncogene theory, we all are living with a sort of time bomb such that at any given moment any one of us will develop cancer. We are all programmed to die. This oncogene theory is pure junk.

Molecular biologists are obsessed with linking all human activities, traits and diseases to genes, even ones that do not exist. Molecular biologists in cancer research hunt for phantom oncogenes. Their experiments are very expensive and produce radioactive wastes. They butcher countless animals to collect genetic material. They are spending billions of dollars each year pursuing a fantasy. Deplorably, it is almost impossible to get a research grant unless a search for oncogenes is included in the proposal. It helps considerably to have a molecular biologist included in the project at an appropriately high position as well. This is one more way that molecular biologists impose their heavy hand in misdirecting cancer research.

Chemists' Causes

Chemists have a favorite truism: Chemists can do biology, but biologists can not do chemistry. The reason for this is simple. Chemistry requires a mind highly skilled in deductive and inductive reasoning. Chemists learn scientific principles and apply those principles to a situation. Biologists merely memorize facts and indulge in pattern recognition. It is no surprise that in cancer research, dominated by biologists, including physicians and molecular biologists, no great advances have been made. Molecular biologists discovered how to spell molecular, therefore they lay claim to and expertise in the understanding of chemicals. Chemicals are made of many molecules. But, molecular biologists refuse to even consider examining any biomolecules other than DNA and, secondarily, RNA, the two molecules coding for genes.

If you were to look at the history of cancer research, you would find that all of the major advances were made by chemists. Chemists developed the techniques that molecular biologists try to use effectively. Unfortunately, chemists also conduct useless experiments in cancer research that offer no new knowledge. Physicians and biologists have an excuse because they are not the brightest boys and girls to walk the planet. If physicians and biologists could think, they would become chemists. Chemists have no excuse for being stupid and incompetent. That they often are is probably attributed to a weak graduate program plus the browbeating by molecular biologists they allow to continue.

Among the achievements of chemists was the isolation and determination of the chemical structures of many carcinogenic compounds. Chemists were instrumental in proving, for example, that several chemicals found in tobacco products cause cancer. Chemists also were able to demonstrate that most chemicals are not of themselves carcinogenic, but that they have to be metabolized by the body into an "activated" compound, which is the true carcinogen. Chemists have made some strides toward understanding what structural features of chemicals are required for a chemical to have carcinogenic potential.

Chemists have as one major goal the development of structure-activity relationships for chemicals in order to determine which have the potential to be carcinogenic. This means that chemists want to be able to predict the biological activity of a chemical based upon its structure. It is known generally that if the structure of a chemical is slightly changed, then the response by the body, animal or human, will also change. To date, however, the goal of deriving structure-activity relationships has not been achieved satisfactorily. This would be a tremendously important breakthrough with enormous benefit. If this goal were achieved, chemists could determine which chemicals should be pulled from the market as carcinogens. In the design of new drugs and other chemicals with human exposure, chemists could examine the new structures and know which compounds not to synthesize and test. Human lives and billions of dollars could be saved.

Why haven't chemists been able to derive structure-activity relationships? Again we have to examine the types of experiments they perform to find the answer. Remember, as discussed previously, cancer does not form indiscriminately everywhere inside the body when the animal is dosed with a carcinogen. Chemical carcinogens target different organs depending upon the structure of the chemical. For example, I worked with a very potent chemical carcinogen found in the side-stream smoke of cigarettes, cigars and pipes. This chemical is called diethylnitrosamine or DEN for short. When DEN is given to the male Syrian golden hamster, lung tumors are formed almost exclusively. When DEN is given to the male rat, tumors are usually found only in the liver; lung tumors do not form. If the structure of the chemical is altered just slightly to make dimethylnitrosamine or DMN, another carcinogen found in tobacco products, then liver tumors form in the hamster, with no lung tumors, and lung tumors form in the rat, with no liver tumors. There is a definite pattern of tumor formation in each animal based upon the chemical structure of the carcinogen. The animals die from the cancer because the tumor cells replace the normal cells, rendering that organ less capable of functioning in the manner it was designed to. The cancer does not spread throughout the animal to every organ, as you may have imagined. Metastasis does not occur, as is currently being taught. This will be discussed later in this book.

Biochemists' Bile

Biochemists like to perform their own brand of useless experiments. One of the experiments biochemists like to perform is to examine the way the liver of an animal metabolizes the carcinogen. They want to know which enzymes are responsible for the transformation of the chemical into its corresponding "activated" metabolite. The rationale for these experiments is based on logic. With an understanding of how the actual carcinogen is formed, researchers can either work on methods to prevent this transformation, and hence the formation of cancer, or researchers can more effectively develop new treatments for the disease.

The goals and logic of the biochemists are unassailable, but there is a major problem with their work. As stated above, chemical carcinogens target various tissues depending upon their structure. What good would it do to study how the liver of the hamster metabolizes DEN when it is not a liver carcinogen? If we accept the hypothesis that an activated metabolite must be formed to produce tumors, it is clear that the hamster liver is not producing the activated metabolite because no tumors formed there. The same reasoning applies to DMN and the rat because no tumors formed in the rat liver. It would seem more logical to study how the targeted organ metabolizes the carcinogen. The possibility that the same set of enzymes are not present in both of these organs is one reason for the difference in the pattern of tumor formation. I was, in fact, able to demonstrate this.

I asked one biochemist why most biochemists used the rat liver to try to understand the metabolism of chemical carcinogens and its relationship to the initiation of tumors. The biochemist replied simply that the liver is used because it is the easiest organ to work with. Perhaps I missed something when I entered cancer research. I was laboring under the illusion that we were suppose to do that work which was required to unravel the mechanism of cancer, however difficult the task may be. It did not take me long to realize that I was violating a taboo of biochemists by not wanting to work with the liver. I worked with the organs that a chemical carcinogen targeted. This was just one of the reasons I was never allowed to publish most of my work. Millions of dollars in pointless research is still being conducted only because the liver is the easiest organ to work with. None of that work can add a greater understanding of cancer. Conclusions drawn from data gathered from the liver go into the development of theories about the mechanism of cancer that may not have any applicability to other organs.

Biochemists have a favorite compound to use when studying the metabolism of any chemical by the liver of any animal. This compound is called piperonyl butoxide or PBO. PBO temporarily inhibits the activity of a group of enzymes called the P-450's. This is the darling enzyme system of biochemists. It is found in large quantities in the liver of all animals. It is a major enzyme system that detoxifies chemicals we ingest, including drugs. It is logical for biochemists to examine this enzyme system for its potential to form the activated carcinogen in the liver. As you might have already guessed, it also is relatively easy to isolate from the liver. Books have been written about how to extract P-450's from the liver, describing in detail various experimental methods using them.

Here is a problem I posed to some biochemists. What happens when the target organ of the chemical carcinogen does not contain the P-450 enzyme system, but tumors form in that organ? How is the activated metabolite formed? I got no real answer from any of the biochemists but I did get a response. The response came in the form of suppression, which is another reason most of my work has not been published. I had inadvertently bruised their egos. I asked them an honest and a legitimate question, which they interpreted as a confrontation. I was branded a troublemaker for simply asking.

In an effort to determine what other enzyme systems might be involved in target organs other than the liver, I applied techniques learned in pharmacology. These techniques can be somewhat complex, but they are well documented in pharmacology textbooks. Pharmaceutical companies must use these techniques in order to have the FDA approve any new drug for the market. I took the theoretical position of treating chemical carcinogens as drugs. This is a sensible approach because drugs are chemicals, just like carcinogens. One of the techniques pharmacologists use to determine the enzyme system they may be dealing with is to use a known inhibitor of that system. If the metabolism of the compound is inhibited (blocked), then it is safe to assume that this enzyme system metabolizes the chemical. By using a set of enzyme inhibitors, I was able to demonstrate that enzyme systems other than the P-450's were responsible for forming the activated metabolite. The biochemists decided my work was too controversial. I asked why. They told me it was because I selected inhibitors which never had been used in cancer research before, and that the enzyme systems I named were never viewed as being involved with the metabolism of carcinogens. I was not allowed to publish the results of my experiments. I could routinely reproduce the results of my experiments. Anyone who followed the procedures I outlined also could obtain the same results. The quality of the work was not questioned. My work was controversial only because I used chemicals never tried before in cancer research to demonstrate that other enzyme systems besides the P-450's were involved in the formation of tumors. Criticism of my work by the biochemists also arose from their ignorance of pharmacology and the techniques routinely used by pharmacologists. I suffered the same retribution as Lake and Cottrell had earlier.

Pharmacology's Utility

There are very few pharmacologists in cancer research. Those recruited to cancer research are there because current theories about all aspects of cancer are failing to answer the questions and the heads of programs have to bring in new ideas to keep their programs running. There is pressure on directors to answer questions or face the elimination of their programs. In an act of desperation, they are recruiting pharmacologists. This is one of the reasons I was selected to work at Bionetics Research, Inc. As expected, however, these same directors are trying to claim the ideas presented by the pharmacologists as their own.

One important concept I introduced and successfully demonstrated early in my post-doctoral career was what pharmacologists call selective uptake. Selective uptake means that a drug is preferentially taken in from the bloodstream by a target organ due to a very specific receptor on the membrane of the cell. More accurately, the selective uptake mechanism refers to cell types rather than whole organs because the same receptor can be located throughout the body. For example, when you take a pain reliever like aspirin, it is selectively absorbed by the nerves that sense pain as opposed to the nerves that move your muscles.

When I worked at the University of Tennessee Veterinary Teaching Hospital, I performed a similar set of experiments with human lung cancer cell lines to demostrate the selective uptake process. My former supervisor would not let me publish all the results of those experiments. But when she toured the world attending cancer conventions, she proudly spoke about this process and presented it publicly as her own idea. However, she spoke of a different type of receptor than I had reported, using those same cell lines. Her results are questionable at best because one can not casually change the type of receptor like one can change shoes. Receptors are complex proteins designed to interact with very specific chemical messengers. Different types of receptors represent different lines of chemical communications. It would be comparable to trying to plug a toaster requiring 120 volts into a 220 volt socket. It just won't work.

This supervisor was able to pull off this scam for two reasons. One, she had the name, which meant that she had more clout than I did. I already discussed how supervisors steal ideas from subordinates. Two, no one else in cancer research either had ever heard of pharmacological receptors or they did not understand the utility of the theory. Therefore, there was no one to challenge the inconsistencies in her presentation. In fact, some researchers heralded the idea she presented as a major breakthrough in cancer. My name got swept under the rug, so to speak.

My supervisor at Bionetics told me he was not convinced of the selective uptake process in target organs of chemical carcinogens. I showed him my data. The results were reproducible. I had less experimental error than is typical with experiments conducted on animals. Yet I was not allowed to publish these results either. Nor was I even allowed to publically present this work to colleagues.

About a year after I showed him my data, another group of cancer researchers published an article demonstrating the selective uptake of a chemical carcinogen that caused pancreatic tumors in hamsters but not in rats. One of the authors of that article had worked in the same laboratory I had, but had left years prior to my joining the company. And he had reported to the same supervisor at that time. This was the same supervisor who would not let me publish my work demostrating the selective uptake process and its relationship to tumor formation. Here is a second example of an idea being stolen by a supervisor. I had shown the supervisor my data more than a year before that article came out. In addition, I routinely discussed the results of my experiments with him. It is my belief that this supervisor communicated my ideas and results to the other group of researchers, for whatever his reasons, to block the advancement of my career.

Since I left both supervisors, neither has been able to continue this work properly. Here is an example based on personal experience of how the thief can not progress with the stolen idea any further than the limits of her or his understanding. Also, both supervisors and the research group are good friends and constantly keep in touch with each other. They comprise part of one of the cartels I have discussed. Because I had been a member in this cartel, due to the petty fueding, it is virtually impossible for me to go to another cartel and have my work published.

While I was at Bionetics Research, Inc., the company had a sight visit by a committee made up of other noted cancer researchers from different facilities. It is the function of such committees to evaluate the quality and necessity of the work being conducted by the facility they visit. All of the Ph.D.'s are suppose to present their work to date and to discuss future projects. Even though I had introduced the concept of pharmacological receptor involvement in the selective uptake of carcinogens as the mechanism by specific target organs, my supervisor did not let me present my work to this committee. He presented portions of my concept. Of all the researchers who talked to this committee, the only idea the committee was very enthusiastic about was my receptor idea. In fairness, he had thought of the concept of receptors being involved in the targeting of organs independently of me, but he never was able to define what those receptors were nor to design experiments to gather data as support for this idea. The experiments he presented to the committee as attempts to define those receptors would not have been successful. His experiments were designed to verify that chemical carcinogens bind to proteins. While it is true that all pharmacological receptors are proteins, it is not true that all proteins are receptors. Therefore, demonstrating that chemical carcinogens bind to proteins does not prove that they bind to specific receptors. The results of his experiments would document the obvious without adding insight to the specific. I, on the other hand, not only had precise and correct definitions of which receptors were involved for several carcinogens, I already had publishable data to verify my concept before the committee arrived.

A Basic Mess

I am certainly not the only one victimized by the theft of scientific ideas, or by the stifling of concepts from innovative minds. These fascist tactics are experienced by every creative scientist. The reality of scientific research is that only a small handful of individuals are capable of generating truly original ideas. This is not surprising or unexpected in any field. Look at the performance of professional athletes. Only a very few have the talent to become superstars. While it is impossible to steal talent from an athlete, however, it is very easy to steal ideas from a scientist, especially from one at the lower end of the hierarchy. When the pressure is on program directors with limited talent to produce results, they resort to stealing other scientists' ideas.

If it appears that basic cancer research is a chaotic mess, this is simply because it is. There is no real cooperation. Program directors can even be possessive about the chemical carcinogens they are working with. One of my previous supervisors ordered me not to work with two chemicals she was interested in even though I laid the groundwork for her series of experiments. A section head who worked for the NCI at Frederick, Maryland tried to prevent me from working with another chemical carcinogen because he had been examining this compound for about 20 years. Members within research cartels cooperate each other, but even here the lines are carefully drawn and maintained.

Physicians, molecular biologists and chemists are fighting each other in the race to claim the theory that will explain the mechanism of cancer. It's time to put an end to this madness. Those individuals had their opportunities in the laboratory and they failed miserably. It is time for the old guard to be purged to make way for young, talented scientists who can work without the threats of the current terrorist regime. The old guard have squandered our money too long. They have built up our hopes far too often, only to invariably crush them.

CHAPTER 12

WITCH HUNTS

Many of you are rightfully confused about just what cancer researchers are doing. We all have heard reports about a particular chemical causing cancer, and then a few days later another report refuting the findings of the first. If these conflicting reports happened only once in a while, then perhaps we could take them seriously. However, this sensationalistic style of alleged science occurs far too frequently. The result is that there is a tremendous credibility gap between cancer researchers and the public. You are justified in being skeptical about the work going on in cancer. Let me explain how and why this unprofessional method of conducting and reporting scientific information is proliferating at your expense.

The Hunters

To begin, virtually all of these reports come from physicians. I have already explained why this group are not scientists, but only know some scientific terms. This reason alone accounts for much of the shabby research you are purchasing. The news media is also to blame. Have you ever noticed that almost all of the medical science stories in the media come from either the Journal of the American Medical Association (JAMA) or from the New England Journal of Medicine? This rubbish is reported as though it were gospel. These magazines compare to responsible scientific journals much the same way that the supermarket tabloids compare to the New York Times. To find real answers to your questions about cancer, you would have to read scientific journals like Biochemical Pharmacology or the Journal of Medicinal Chemistry. A major drawback to understanding articles in these journals is that it requires some knowledge and education in the area the article addresses. Perhaps this is one reason the news media never reports information from these sources. They do not have people who can understand this scientific literature.

There is another reason we hear conflicting reports about virtually every topic in cancer research. Most of you have heard the adage that in research one either publishes or perishes. This is true for the most part. Almost all cancer researchers feel this pressure to publish. However, very few have sufficiently creative minds to develop new concepts worthy of investigation. Most cancer researchers resort to one of two tactics: 1. They steal ideas from other researchers, or 2. They go on a witch hunt. The witches they hunt are anything they can think of that might be forced into a relationship with cancer.

This is exactly the same mentality that was present in the Middle Ages when the Spanish Inquisition roamed Europe. The Inquisition was a quick and easy way for a priest to rise from obscurity to prominence within the ranks of the Catholic Church. In efforts to gain property and wealth for the Church, those priests charged nobility who did pay enough money to the local parish and to the pope with being possesed by the devil. Lesser ranked individuals also were maliciously harassed for the sake of religion. Young virgin girls were their favorite target. After several hundred years of this horseplay, enough people became disgruntled and put an end to these activities. The methodology for a cheap and dirty way to rise to power had become established.

About 150 years ago, men of science took over the job of hunting witches. This time their target was God and religion. Following the publication of Darwin's theory of evolution, the scientists believed they could solve all of the world's problems, and set out on a course to ridicule all religious beliefs. This was one of the guiding forces behind Karl Marx and his Communist Manifesto. As we now know, scientists do not have substantially better solutions to world problems than the rest of us. But the burning desire by some to cause trouble while claiming to do good for humankind was and is still very much present. God is not the popular target He once was. Because torturing, raping and burning at the stake of virgins became likewise passe, marauding scientists had to find another target for their vicious assaults. About 50 years ago their dilemma was solved. They discovered their witches in chemicals, viruses, and genes.

Easy Prey

Carcinogen witch hunts are very easy to conceptualize and to perform. All any researcher has to do is simply pick a chemical at random and claim it is a suspect carcinogen. If he chooses not to handle chemicals, then that researcher can pick mechanical devices such as high voltage wires or silicone breast implants. Lifestyles are included in the hit list. These researchers, most often physicians, use the little imagination they have to find targets for their witch hunts. And they invariably use statistics to prove their point. This is why you hear such drivel as, "Researchers have found that children who live near high voltage lines have a 1.7 greater chance of developing cancer." When these researchers try to burn their witch at the stake by branding an object a carcinogen, there is also someone waiting to "rescue" the witch by conducting experiments to refute the initial report.

Physicians and biologists like to focus their attention primarily on chemical witches. Recently, nevertheless, they have teamed up with molecular biologists to find a non-existent entity they call the oncogene. Molecular biologists are spending billions of our dollars and polluting our environment with radioactive wastes from their experiments to hunt these oncogene witches. Chemists also try to find new witches, usually chemicals in the form of carcinogens so that they can make grandiose claims for highly toxic compounds they have developed and arbitrarily labelled as cures for the disease.

What is the goal of these researchers? It is not to protect anyone from cancer. It is a cheap trick to make a quick name for themselves in the field of medical research. Until recently, it was a guaranteed publication to mindlessly report a chemical as a carcinogen. That researcher's name got into the press, he became more visible to colleagues, and a credit to his supervisor. This was standard procedure because it is easy to do and very difficult to prove faulty.

The limited minds that run cancer research programs are not gifted investigators either. A new person in the field who displays similarly limited talent will not be a threat to the director of the program. As a consequence, that person is hired over a talented scientist. Eventually the director retires, with his minion taking over as head of the program. This new mental midget continues the practice of hiring inferior investigators in efforts to cover up his own inadequacies. The cycle goes on for as long as they can all milk the system.

That one researcher refutes the findings of another does not indicate that he is particularly bright either. It does not require creative genius to ram chemicals into rats and wait until they die of cancer or the chemical. It also does not take talent to ram a chemical into a rat and report that nothing happened. Neither researcher can explain why their respective results were obtained. Both have accomplished their mutual goal of making a name for themselves as quickly as possible. It is not too surprising that researchers on opposite sides of the issue are friends. By maintaining their dialogue through the reporting of opposing experimental results, each researcher is presenting to their supervisors a measurable degree of activity. They maintain a high profile with the public as well. You can consider these reports as a show of force by the cancer research community to impress the public.

These con artists have been able to get away with their witch hunts because you are not aware of their tactics, or that better investigators exist. Some of the best scientific minds are no longer in research because they have been systematically targeted for eradication by the witch hunters, who see them as an obvious threat to their scam.

Statistical Weapons

One tool of deception is the use of statistics. Witch hunters commonly apply statistical analysis to both basic and applied cancer research. A typical story in the news goes something like: "Researchers have shown that women who smoke past the age of 40 have a five times greater chance of developing breast cancer." This statistic may or may not have applicability to the general public, although it is used to prove that cigarette smoking is bad for us. *If a scientist needs statistics to prove a point, then that researcher has no point to prove.* Statistics are a valuable research tool if properly applied. The proper application of statistics in medical research is only to indicate any trends or patterns that may be apparent in the data. To prove that smoking is bad for women over 40, a direct cause and effect relationship must be consistently demonstrated. In other words, the researcher must dose women over 40 with tobacco smoke and show a 100% incidence of tumor formation resulting from that dose. Anyone else who conducts the experiment as originally described also must obtain exactly the same results. Of course we can not perform such experiments on people. We have to rely on other methods, like animal models coupled with statistical analysis. I do personally believe that all tobacco products are bad for everyones' health. I object, however, to the misuse of statistics by those researchers who are out to make a quick name for themselves without further progressing our understanding of the disease.

A Media Event

Have you noticed also that the news media routinely reports something about cancer as if on schedule? These stories often deal with promising new leads for diagnosing, preventing or curing the disesase. These press releases are circulated so that you get the impression that real progress is being made. Researchers sense your impatience. The public talks to them the same as to me. Legislators are putting pressure on them to produce. Remember, politicians are not immune to cancer. Bogus reports are released every now and then in an attempt to put suspicions to rest, and to ensure an income for at least several more years.

Some researchers specify the type of cancer they are working on. Stories that specify a single cure for a single cancer are based upon a portion of my theory on the mechanism of cancer that I had to divulge to some people in the field. I originated this idea, but without credit. To date, none of the thieves have been actually able to produce any cure because they do not understand my complete theory. This is one more example of how the thieves are incapable of fully implementing the ideas they steal. It also is evidence for my assertion that an investigation into the field of cancer research must be made to rid it of the thieves so that competent scientists can work on the problem.

Another type of press release announces "a promising new treatment". A molecular biologist, for example, may report the location of a new oncogene. The press then exploits the implication that locating this gene offers promising new strategies for fighting cancer. I find it interesting that a mechanism is never described that relates the oncogene location to any ability to fight cancer. You are just expected to be encouraged by these "breakthroughs" so you will keep the tax money and donations coming.

In the classic shell game, all you have to do is find the ball under one of three shells. The shells are moved quickly around and you pick the one the ball is under. If you pick the right shell, you win the prize. Sounds easy enough to do, except that the ball is never under any of the shells when you make your pick. You play because you think you can win. Cancer researchers work the same scam. They report that strides in the field have been made. If only we give them more money and more time, they will find the cure. For most directors of cancer research programs, the dollars are running into the billions and the time expended exceeds 30 years. If you challenge these researchers, you will find that no really new information has resulted from their projects. You have a right to be impatient. You have the right to demand that they produce or get out of the laboratory and let someone in who can. This is how the business is run where you work. The business of cancer research should be conducted in the same manner.

By reporting periodically some useless result and/or bogus data, they give the illusion that a major breakthrough is just ahead. Perhaps some of them honestly believe there is one. But the purpose for which they were hired is not to simply work in the laboratory. They were hired because they claimed they needed the laboratory to produce the results they promised - curing this disease. Even if they have the best of intentions, it is not enough.

A Ghost Hunt

A homeowner may be willing to work hard to get rich. There is nothing wrong with this. So he decides to dig for gold in his backyard in New York City. His intentions are good, but let's face it, he has no real chance of succeeding. He is working on his goal, but all his efforts will be futile. So too with cancer research. The work going on now is just a series of futile attempts by incompetent, unimaginative, egotistical failures. Despite the most vigorous efforts by these witch hunters to link a variety of chemicals to cancer formation, no one to date can look at the structure of a compound and determine that it is a carcinogen. The only way they make this determination is to test the chemical in animals. To understand better how these con artist contrive their research projects, I will describe in detail how you can do cancer research of equal quality right in the privacy of your home.

CHAPTER 13

HOME GROWN CANCER RESEARCH

In previous chapters, I described in detail the types of cancer research being performed and how con artists posing as scientists are getting away with their scams. It is difficult to describe all the subtle nuances of their shell game. By reading this chapter you will learn how to conduct cancer research inexpensively at home that will be of equal value to that reported by the news media. View this chapter as a fun exercise. Think of it as being comparable to a handbook on magic in which a magician reveals some tricks of the trade. That is the purpose of this chapter, to reveal the tricks of the trade in cancer research. You will have acquired first-hand knowledge of how easy it is for cancer researchers to present their misguided notions.

The first thing you need is to decide the type of cancer research you wish to conduct. You have the choice, according to your tastes, between the two broad categories of basic or applied research. Applied cancer research is the more expensive and difficult of the two, but, once you learn the techniques, it is relatively easy. One caution about applied research is the decision to use mice or any other traditional laboratory animals. Remember, the initial mouse will cost you about $150,000. Additional mice, of course, will be somewhat cheaper. Of the two types of cancer research, basic research is the cheapest, with the most return on your investment in terms of making a quick name for yourself. For this reason, our home course will focus first on basic cancer research.

Home Grown Basic Research

To conduct basic cancer research at home, the only major piece of equipment you need is a personal computer, or PC. If you already have a PC at home this will not be an additional expense. For the PC you will need two programs. These are a word processor and a statistics program. The word processor program makes writing articles easy, while the statistics program is needed to analyze the data you will generate. Even if you never had a course in statistics, you can still use the program by simply following the instructions that come with it. The program will even print out a conclusion based on the analysis. The computer will tell you whether or not your data is "statistically significant". You soon will learn how this phrase is used in cancer research. The two programs cost from $200-400 apiece. Perhaps you already own the word processor program. If you do not have a PC, you can expect to pay $1500-2000 for one, and then about another $600 for the two programs you need. Your money will not be wasted should you decide to drop out of cancer research because you can use the PC for other equally worthy purposes, like playing video games.

Being Trendy

In basic research you want to obtain fundamental information, which you can claim to be necessary for unraveling the mechanism of cancer. In order to get this data, you have to go on a witch hunt. When picking a witch, be creative. After all, some cancer researchers recently decided that fluoride is a carcinogen. And you know that if high voltage wires can be a witch, then anything can. Your selection is just as valid and justifiable as any physician's or molecular biologist's choice. Who knows, they may even join you in your hunt. If you want drug dealing out of your neighborhood, link drug trafficking to cancer. If there is a business you want out, link the activities of that business to cancer. Did you buy an inferior product from a manufacturer? Link that product to cancer. The number of possibilities is endless.

You also have the choice to exonerate a witch. You may have a need to prove that something that was once labelled a carcinogen is safe. Again, I can use fluoride as an example. One group called it a carcinogen, while another group proved that it was not one. There is no mechanism by which I can envision fluoride as being a carcinogen. It is highly toxic, but not in the amounts put into toothpastes. So you have these two paths to take in basic research, to accuse or absolve.

After you have selected your witch, you must then choose the type of cancer you wish to link to it. Selection of a cancer can carry an associated political liability. Some selections are safe in that there is current pressure to examine those cancers, such as breast cancer in women. Men can also develop breast cancer, but you do not want to get involved with that controversy. Leukemias, especially in children, are always very popular and politically safe. Colon and prostate cancer are likewise fashionable at this time, but stay away from liver and lung cancer. You will have the weight of the tobacco industry down on your back in no time. You might end up facing costly lawsuits to defend your position. Liver and lung cancer should be left for later, as advanced topics in your cancer research.

To begin your work, start out slowly. Pick a socially acceptable and politically safe cancer like leukemia. Both children and adults develop leukemia, so it elicits a great deal of public interest and empathy. You can exploit this interest for financial support of your work, much like the leukemia organizations of today. Later I will tell you how you too can even start your own cancer organization.

Defining Your Project

To guide you through this course in basic research, I have chosen chocolate cake mix as my witch. It does not matter which variety. Even though I have no data to support my selection and no real reason to believe that any mix is carcinogenic, I will still be able to "prove" any one of them is a carcinogen. I will link these mixes to leukemia. This is an excellent choice for several reasons. First, both children and adults like chocolate cake. I can choose to milk the sympathy of people who favor either group. Second, chocolate cake mix is a relatively complex mixture of many different chemicals derived naturally from plants. Chocolate, for example, is extracted from the cocoa plant. No one has identified all of the chemicals that occur naturally in chocolate. Just isolating all of those chemicals can take a lifetime of work. Chocolate cake mixes also contain either sugar or a synthetic sweetener. I can always link cancer to the ever controversial sugar. The synthetic sweetener is fair game too, as the case against saccharin a few years ago illustrates.

Cake mixes contain flour. Humans have been eating flour since before recorded time. You may wonder how in the world I can link flour to cancer. I have inside information on this product. There is a northern province in mainland China which has a much higher than normal incidence of tumors in the esophagus of women. Analytical chemists have isolated and identified a

150

chemical carcinogen that is produced by a certain type of mold that can be found in wheat from that region. When laboratory rats were dosed with this chemical, tumors of the esophagus formed. Now you have the necessary ammunition to call into question the safety of flour. I bet you didn't think I could do it. Do not be bashful about your selection of a witch, or type of cancer.

We now have a carefully defined cancer research project. Reviewers of grants like to see phrases such as carefully defined. We are going to prove that eating chocolate cake prepared from commercial mixes causes leukemia in children. I chose children in order to play upon the emotions of the public and the reviewers. Any con artist worth his weight in garbage would do the same. We now have all of the necessary ingredients to perform cutting edge style basic cancer research. We have a witch, a type of cancer and a population for statistical analysis. In addition, we can prolong this project indefinitely by testing the multitude of chemicals contained in the chocolate mix. You should be excited because now we are going to really start researching.

Surveying Your Subjects - Selectively

To perform valid statistical analysis, the best approach is to obtain the largest number of subjects. The ideal situation would be an infinitely large number. Practically, we have to settle for a much smaller group. The formulas contained in your statistical program will adjust for this limitation, so don't worry about it. We must define the characteristics different groups of children must possess in order to compare the effects of eating chocolate cake to the development of leukemia. We can define our groups as, Group 1: children who have no leukemia and who have never eaten chocolate cake; Group 2: children who have leukemia but who have never eaten chocolate cake; Group 3: children who have eaten chocolate cake and have no leukemia; and Group 4: children with leukemia who have eaten chocolate cake.

The next step is to prepare a survey to obtain the information we need. The questionaire for this survey can be simple with only two questions. 1. Have you ever eaten chocolate cake made from a commercial mix? 2. Do you have leukemia? If you want to get more involved in the project and statistical analysis, you could ask a third question such as: 3. How often do you eat chocolate cake? With this third question, you can show a correlation between the frequency of eating chocolate cake and leukemia. Cancer researchers will consider your work to be of higher quality if you do so. In addition, you will take away a research project from another researcher who can see the "need" to show that correlation. Include the third question in your survey.

The next step in doing your own research is to code the responses to the questions. Keep it simple by making a yes answer equal to 1 and a no answer equal to 0. All you have to do is type a 1 or 0 into your PC with the statistics package. The computer will do the rest. It will print out what is called the correlation coefficient. This number will be a decimal. The larger the number, the better the correlation. The computer also will tell you if that correlation is statistically significant. Even if you do not understand statistics, the computer programs are designed to analyze the data for you, including the conclusions you should draw. Relax and let the computer do the work.

Finally, you have to select your subjects. Proper statistical methods require a large number of subjects. But, asking every child in the world clearly is impossible. Selecting 200 children should be an adequate sample size. The computer program will make mathematical adjustments to analyze the data as though an infinite number of children were sampled. Now all you have to do is ask 200 children those three questions and place them into one of the four groups listed.

Getting the Right Results

Suppose that after you have completed this experiment no correlation was found between leukemia and children who eat chocolate cake prepared from mixes. What could you do? There are at least three options available. 1. You can bias your sample population by going back out to find more children who have eaten the cake mix and who have leukemia. You then can put this data into the computer. 2. You can just simply not include in the data as many of the children who do not have leukemia. Both of these options will readjust the statistical analysis in your favor. Physicians frequently use both of these options when they either witch hunt or use statistics to evaluate new treatments that they want to succeed. 3. The third option is to lie. Fake your data. A number of prominent cancer researchers make up their data to support their pet theories. Many of the popular concepts concerning cancer are fabrications by such researchers. These concepts are not supported by either scientific data or principles. When you lie about your data, you will be in prestigious company.

No, I am not being cynical. I have personally witnessed the falsification of data, and those in the field have reported to me their witnessing of data fabrication. There was even a special program on my local public television station that dealt with this very issue. It was titled: "Do Scientist Cheat?" It was produced by Rob Wittlesey. I urge anyone who has not seen this program to do so.

I know of one company that was founded by a molecular biologist who made the claim that he had a new technique to modify oncogenes. He was to do something with oncogenes such that tumor cells transformed back to normal cells, and thus cure cancer. He received millions of dollars from private investors to start his company. He presented them with convincing data that offered the ever popular " promising new lead" in cancer treatments. One of the technicians I knew who worked for this molecular biologist's company called me several times to ask for my advice, because as often as she tried to reproduce the molecular biologist's work, she never could. I advised her to leave the company as quickly as she could find another job because I knew from her description of this work that there was no way the reported results could be obtained. The methodology went against the principles of science and sound molecular biology theory. In short, the molecular biologist had falsified his data in order to get funded for his research project. After about three years of operation, his company folded. He inadvertently made a public demonstration of his scam.

Of course, if you want to "prove" that chocolate cake mixes are not carcinogenic, then manipulate your data in that direction. This would be a wise tactic if you are interested in being hired by any company that manufactures cake mixes. You do not have to worry about your report being challenged. It takes time, money and effort to question your results. To obtain the time and money, there has to be some sort of justification presented to those individuals who can fund the effort.

On the other hand, if you link a chocolate cake mix to leukemia in children, the company making that cake mix will have a strong interest in disproving your findings. That company will invest the money to refute you. This was the situation with a solvent organic chemists routinely use called methylene chloride. It was reported as a liver carcinogen in mice. After expensive counter studies by its manufacturer, it was discovered that most of the mice in the original study died from exposure to this chemical long before there was time for any tumors to form. Even then the incidence of tumor formation was low and could not be definitively linked to methylene chloride exposure.

Home Grown Applied Research

You have just had an introduction to basic cancer research. Some of you may be more interested in curing the disease. You still can work in your home on applied cancer research. This will be more expensive than basic research, but the financial rewards can be great. You will need to make some modifications to a room in your home in order to do this type of work. You remember how expensive a mouse can be? You're in luck. You can develop anti-cancer drugs for use in humans without using animals at all. You can use tissue culture of cancer cells. If you can demonstrate to the FDA that you have a compound that kills cancer cells in tissue culture, then there is an excellent chance you will be able to try that compound in humans.

This was done in one of the cancer research facilities at which I worked. The chemists tested their compounds on a variety of cancer cells. Some of the compounds were more potent than others. They received FDA approval to try them in humans. All of the compounds used were variations of rocket fuel. I mean literally rocket fuel. These compounds were highly reactive chemicals that bound rapidly to biomolecules of all kinds. If these compounds were not handled properly, they had the potential to be nearly as explosive as nitroglycerine. But the FDA approved these poisons for use in humans on a trial basis. The important lesson here is to not be shy about using any chemical to demonstrate potential as an anti-cancer drug.

After you complete this exercise on the development of anti-cancer drugs, you will be able to understand better how such bizarre remedies as the Chinese cucumber and the apricot have been touted as anti-cancer treatments. I am not sure of how the Chinese cucumber works, but I do know that apricot pits contain a relatively high concentration of cyanide. You are now ready to conduct applied cancer research using items found in your kitchen.

A Place and Procedures

To do applied cancer research at home, you will need to set a small section of your house apart to become your tissue culture laboratory. The major pieces of required equipment are an incubator @ $8,000; a laminar flow hood @ $6,000; an autoclavc @ $5,000; a microscope for tissue culture @ $1,500, cabinets and counters @ $2,000; a small refrigerator with freezer @ $700; and glassware @ $1,000. Including the cost of other materials and supplies, you have to spend about $25,000 on equipment. Renovations to your laboratory to install the equipment will be about $5,000. For a total of about $30,000, you can do applied cancer research at home. The quality of your work will now be equal to that of any chemist currently involved in applied cancer research for a pharmaceutical company, if you follow my instructions.

You can buy a large variety of cancer cell lines. These cancer cells come from mice, rats, hamsters and even humans. It does not matter which types of cells you choose because all of them are grown using the same technique. To get a copy of catalogues for the equipment and cell lines, just go to your nearest medical school library and ask the librarian. You buy cancer cells exactly the same way you get tires from a Sears catalogue.

Growing cells in tissue culture is a relatively simple technique. The most important thing to remember is to keep your laboratory clean and neat. You should have a supply of isopropyl alcohol in a spray bottle on hand at all times so that you can sterilize counter tops and the inside of the hood. There are textbooks in the library that give detailed instructions on tissue culture techniques. You buy about one quart of the prepared tissue culture medium. The autoclave is used to sterilize the glassware. The cells come frozen. They have to be gently thawed. The technique for thawing cells is described in the textbook. Once the cells are thawed, they are placed into specially designed plastic bottles. You add about two ounces of culture medium to each bottle and then place it in the incubator. The incubator has to be set at 98.6° F, or 37° C, with 100% humidity. The cancer cells will begin to multiply shortly. You should have at least 10 flasks of cells for each compound you want to test. Within about two weeks, you will have enough cells to experiment on.

One essential procedure is to take a portion of the cells you have grown and use it to seed other flasks. This way you will keep the lines growing constantly without buying new cells. Do not worry about handling the cells. Cancer is not infectious. You can not "catch" cancer from them. The cells can not live for very long outside of the carefully controlled environment you are placing them in. If any cells come into contact with your skin, they most likely will die. You should wear plastic gloves and keep the environment as sterile as possible to prevent micro-organisms such as yeast and bacteria from invading the cells.

Once you feel comfortable with the technique of tissue culture, you are ready to conduct experiments. I suggest you buy at least six different cancer cell lines. You will want to demonstrate to other researchers that your compounds preferentially kill certain cell types over others. If you can demonstrate this selectivity, you will earn the interest and respect of other researchers.

Recipe for Success

Next, let's pick four white powders in your kitchen. I have chosen white powders because physicians and molecular biologists believe white powders are more pure than colored powders. We can choose salt, sugar, flour and baking soda to test as anti-cancer drugs.

After you select your white powders, you can merely dump into each flask any amount you want. But to be scientific about it, you need to measure the quantity placed into each flask. You can use tablespoons, or to be even more scientific, you can buy a metric balance for about $1,500 to get precise weights. The experiment should be conducted as follows.

From a flask containing a cell line, dilute the number of cells and place this dilution into other labelled flasks. The labels should contain the cell line, date and quantity of white powder added. Record in your notebook this data along with the number of cells each flask contains. Detailed techniques for cell counting also are described in the textbooks. Once you have added the powders, count the cells in each flask the next day. You should see a decrease in the number of viable cells. You can tell if a cell is viable because it will adhere to the bottom of the flask. Dead cells are spherical and they float. When you rinse out the flask, the dead cells will wash away with the spent medium. On graph paper, plot the amount of powder used versus the number of viable cells. This plot will give you what is termed a dose-response relationship. This term means that as the quantity of powder is increased, the number of viable cells will decrease.

You may not be able to kill any cells with the first doses you try. Do not be discouraged. This happens to scientists on the cutting edge of research. Should this result occur, simply add more powder. I guarantee there is some quantity of each powder that will kill the cells. You also will find that each cell line responds differently to each of the powders selected. This is a desired response because now you can report, for example, that baking soda is preferentially effective against a mouse leukemia cell line than is salt.

Now you can write up your results in a paper and submit it for publication. To continue with our example, you can report that baking soda kills leukemia cells. You can then go around the country as an "expert" selling your idea to anyone who will listen. The important point here is that you will not be lying. By the way you conducted the experiment, you were able to show that baking soda is effective against leukemia. You can prepare a press release stating that you have a promising new treatment for this scourge. Who knows, maybe some Hollywood celebrity will support your work with fund raising balls or concerts.

In all probability you will not be able to publish your results in any of the currently fashionable medical journals because you do not belong to a cartel. Physicians, biologists and chemists like to protect their turf from intruders who did not go through the initiation ceremonies of graduate school and post-doctoral experience. This is only a minor detail. By-pass the system. Start a journal of your own. You can make it a community project. Invite your neighbors and friends to try other powders. They can even grind up plants found around the neighborhood, like poison ivy and pine needles. All you have to do is dry the plants and crush them to make a powder. You can add these powders to the flasks the same way you added the white powders. All of you can then report in your cancer research journal that, for instance, poison ivy kills cancer cells. Be sure to try your garden cucumber because in a high enough concentration it will kill some type of cancer. Each person in the neighborhood can write about a different powder or plant extract for your journal on homegrown cancer research. These articles will be just as valid as any others currently being published.

As you gain momentum, you can branch out to other communities as well. In almost no time, you will be able to form some type of cancer research association. You can call this organization by any name other than those currently being used. To raise funds for your research, send your children to street corners with plastic cups that have the name of your organization on it. The children can panhandle for money the same way that on weekends college students ask for money at busy intersections for other cancer organizations. Or better yet, hire a professional fund raising outfit. In time you will not only obtain enough money to support your research at home, but you also will get enough money to sponsor your trips to conventions. You too will be able to vacation, at other people's expense, with the alleged purpose of sharing knowledge with your colleagues. In other words, you can work the same scam as reputable cancer researchers. Don't forget to give yourself a hefty salary from the contributions.

It is from this semblance of scientific investigation that you hear constantly about how something leads to cancer, or of a promising new treatment for cancer. If you report your results as I described, you will not be lying. Cancer researchers may not always be lying either, but their results offer no hope over the results you can report. It is easy for them to manufacture any result they desire because there is no one to challenge them. You will not be challenged either.

Welcome to the Club

Now you can understand how research offering false hope is being constantly reported by the news media. For the record, I can explain the mechanism by which the four white powders I have suggested will kill cells. Notice I did not restrict the mechanism to just cancer cells. Each of these powders will dessicate the cells, but different quantities will be required for each powder and for each cell line. The powders will draw out the water from the cells. Once the water is gone, the cells will die. The problem with using these powders as anti-cancer drugs is that they will draw out the water from normal cells as well. Just like the anti-cancer drugs currently being used, these powders can not differentiate between normal and cancer cells.

Cancer drugs kill indiscriminately just like the white powders I selected. All current anti-cancer treatments are based upon the wishful thinking of those who develop and sell them. If you try to challenge any of these individuals about their products, they will use any tactic to intimidate you into believing they are very near to a cure. The truth is that they have no real understanding of how to rationally cure cancer. You can work in your home on projects with the same quality as those of the cancer researchers you are paying top dollar. Your work will be equally valid, and you can experience first hand how easy it is to fool the public when it comes to cancer research, whether basic or applied.

CHAPTER **14**

THE CELL

In order for you to understand the following chapters, you need to know a little cell biology. I had to make a few changes in the theories on the functions and structures of cells to unravel the mechanism of cancer. I adjusted my understanding of cellular biochemistry based on facts and not on untested suppositions. Knowledge of the structure and various functions of the cell is fundamental to understanding how a normal cell becomes transformed into a cancerous cell. Every cancer starts with the transformation of a single cell type. This concept of cell type is fundamental. By examining the changes that occur in a cell type, one can then make sense out of the confusion surrounding cancer.

A human being is made up of various organs, all of which work together as a unified whole. Each organ is made up of several cell types. In other words, in dealing with cancers, you can not simply refer to an organ. You must refer to the cell types that comprise an organ. Your house in similar fashion, is made from a number of different construction materials. The building has wood, concrete, plastic, glass, metals and other materials combined to make a house. You can think of each organ in your body as being an individual house. Various materials, or cell types, go into the construction of each.

A Genetic Fortress

Our bodies are equipped with a tremendously powerful immune system. We do get sick occasionally, but when you consider how often we are exposed to many different microorganisms without becoming ill it is clear that our immune system works well. Our bodies posess another important defense system which is often overlooked when discussing cancer. This system is comprised of enzymes located within the individual cells, especially in the liver. These enzymes are responsible for destroying chemicals, or for at least rendering chemicals less toxic. These two defense systems, immune and enzyme, work in conjunction to protect our bodies from invasion by foreign matter, be it a microorganism or a chemical.

The structure and function of a cell can be compared to a medieval castle. A castle has an outer wall with gates to allow people to enter the grounds. Many times the wall is surrounded by a moat or the castle is double walled to make entrance even more difficult. This outer line of defense is designed to selectively let people in and to keep invaders out. A castle was self-sufficient in that all of the artisans needed to make weapons, clothing and furniture were located within its walls.

The castle was designed to protect the lord and his family from invaders. As you went further into the grounds of the castle, there were additional defense systems, such as additional walls, moats and escape tunnels. The most central and safest location within the castle was reserved for the living chambers of the ruling family. Cells are the castles for our genetic material.

Cells have an outer double wall made of materials that include protein that act as "gates", which pharmacologists call receptors. Receptors are designed to keep foreign chemicals and microorganisms out of the cell. Receptors are also responsible for allowing only certain chemicals to enter into the cell. These chemicals are required nutrients, such as oxygen, water, amino acids, electrolytes, and chemical messengers like hormones.

Structures inside of the cell make biomolecules that the cell needs. There are enzymes that use chemicals, like glucose, for fuel to enable other chemical reactions to occur within the cell. You can consider the genetic material, called DNA, as the "lord" of the cell. The DNA is housed within the innermost structure, called the nucleus. The nucleus has its own double wall that includes proteins that act as gates for protection.

Like the body as a whole, there are two defense systems that are responsible for attacking and destroying invading microorganisms and chemicals. One of these is the lysosomes. Lysosomes are tiny sacs containing very powerful enzymes that destroy microorganisms by literally chomping them into little pieces. Once these enzymes are released they can destroy the cell itself unless they are stopped. They are stopped by destroying themselves. The second defense system is another set of enzymes. These are located along channels within the cell and are called the P-450 enzymes. Under the high magnification of the electron microscope these channels look very much like a series of canals. The enzymes located here are the ones biochemists usually use when conducting experiments on the metabolism of chemical carcinogens. Each cell contains elaborate systems working together to protect the most important biomolecule in the body, the DNA.

The DNA Library

We can consider the DNA molecule as a library of blueprints for the construction of proteins. The sequence of events to produce proteins is simple. DNA codes for RNA. RNA codes for proteins. Proteins are the work horses within the cell. Proteins "read" the DNA sequence to assemble the RNA. Another set of proteins "reads" the corresponding RNA sequence to assemble yet other proteins. Some proteins, which are enzymes, use glucose and amino acids as raw materials to assemble other proteins for structural support in addition to constructing DNA and RNA. Other enzymes use these raw materials as fuel to provide energy for chemical reactions within the cell. Within each cell there is a complex series of chemical events occurring at a rate of hundreds of reactions per hour that must be carefully coordinated. Proteins are responsible for regulating these events.

One very important function of some cell types is to manufacture chemical messengers that are released into the blood stream. These chemical messengers are hormones and a group of chemicals called autocoids that are responsible for initiating specific activities in other organs. Adrenalin is an example of an autocoid. Increased levels of adrenalin occur when you are frightened or excited. Adrenalin is manufactured primarily in the adrenal gland. From this gland the adrenalin travels to the heart and lungs to increase your heart rate and respiration. We all have experienced this reaction. There are a number of autocoids that allow organs to communicate with each other via the bloodstream. The chemical messengers our bodies manufacture enable us to move our muscles, digest our food, and breathe, among other functions. Enzymes make these chemical messengers.

Unlike molecular biologists, I ascribe a relatively minor role to the importance of the DNA molecule. They believe that the DNA molecule somehow sends out messages to other biomolecules to initiate biochemical events. This thinking is false. As stated above, the DNA molecule is only a library that contains necessary information, but it does not send out any communications to the other biomolecules. DNA just sits inside of the nucleus until a chemical message is received. Once the message is recieved, a series of chemical steps for gene expression is put into motion. If there is no chemical messenger present, then no gene goes into action, i.e., is expressed.

Perhaps you can better understand this concept if you think of how a library functions. When you want to take out a book, the library building does not call you on the phone to let you know that you want it. The building does not come to you to give you the book. You decide that you want a book, and then go to the library to get it. Each section in the library can be thought of as representing a different set of genes. A book on a particular topic, sailing for instance, can be considered analogous to a specific gene. If you want a book on sailing, you must go to the section that has books on that subject. Enzymes involved in gene expression are like librarians. Librarians can guide you to the section where the books on sailing are located.

You are like the chemical messenger. You go to the library and set into motion a series of events to obtain the book you want. You may not know where the right section is in the library. The librarian, like the enzyme, assists in locating that section, or set of genes. You find the specific book, or gene, that you want and express your desire by taking the book to another librarian, or enzyme, who stamps it for you so that you can take it home. Like a chemical messenger, you fulfilled your mission of getting a specific book on sailing and you leave the library. Your desire was expressed. This example is a simplified explanation of how genes are expressed. Keep in mind these analogies of an organ as a house, a cell as a castle and the DNA as a library to guide you in understanding my explanation of the mechanism of cancer.

CHAPTER 15

A BRIEF HISTORY OF THEORIES ABOUT CANCER

Over the years, cancer research has embraced a number of fads. These included theories on the mechanism of cancer, cures for cancer, and most recently ways to prevent cancer. You would not suspect that scientists are prone to embracing fads, given their image of aloof objectivity. Scientific research of any kind is supposed to be based upon sound evidence and basic principles. This, unfortunately, does not necessarily apply in cancer research. Too often, egotistical desires are reflected in the theories cancer researchers develop. Almost all of the popular theories related to cancer have very little or no basis in fact. Much of the data has been manufactured by the researchers in order to support their given pet theories, but you know how easy it is to falsify data.

Let me walk you through the confusions and myths surrounding the propaganda. This chapter is not an all inclusive list of every bizarre theory ever proposed. The focus is on theories that are generally accepted by researchers and the unsuspecting public, especially those that have been perpetuated at the expense of your money and lives.

A BRIEF HISTORY OF THEORIES ABOUT CANCER

Early Research Finding Causes

In the beginning of cancer research there was little relevant information in any area of medically related work. Early theories about this disease can not be harshly criticized, due to simple lack of knowledge. The instruments available were not nearly as sophisticated and accurate as the ones now used. Today, the fundamental problem is researchers with lack of intellectual breadth. Technical specialization has created scientists who are either too conceited about their own field to be bothered to extend their knowledge into other fields, or who are intellectually incapable of grasping knowledge outside of their speciality. The latter case pertains especially to physicians and molecular biologists.

As a result of limited minds examining the problems associated with cancer, researchers have internalized popular political and social ideologies into their theories. This popularist mentality is especially evident in theories about the causes of cancer and theories about how to avoid cancer. But it also is evident in some of the treatment strategies, like using interleukins and interferons. Neither of these extremely toxic proteins are anti-cancer agents. These proteins kill anything they contact. Let us take a closer look at some of the myths surrounding the causes of cancer.

Fifty years ago, about all that was known about cancer was that the disease existed and people died from it. It was logical to examine any possibilities that might offer some clues to the development of the disease. Many factors were considered, including diet, heredity, infection from microorganisms and chemical exposure. After about 30 years of work, it became clear that certain chemicals definitely cause cancer. Some of the carcinogens found in tobacco products, for example, have been widely publicized. Arguments of the tobacco companies notwithstanding, there is no reasonable doubt that some chemicals found in tobacco, such as nicotine derviatives, definitely cause cancer. The link between lung cancer and smoking is not a fad.

In the early 1960's, cancer researchers were consistently able to demonstrate that some chemicals cause cancer in laboratory animals. Those chemicals were isolated and identified in the environment. We are exposed to them everyday. One group of chemicals is called polycyclic aromatic hydrocarbons or PAH. Another group of chemical carcinogens is nitrosamines. Nitrosamines are found in the air as a by-product of car exhaust, as are the PAH's, and in tobacco, cosmetics, and food.

It is a well documented fact that the incidence of all cancers increases the closer a population is to a major industrial center. This fact became personally evident to me when I had the opportunity to visit Hungary. My parents are from that country. My father was born and raised in the agricultural southeast corner of the country. My mother was born and raised in the heavily industrialized northwest corner. If you visit the cemeteries in these two regions, as I did to see my relatives' graves, you will find that the people buried where my father came from typically lived into their 70's and 80's, both men and women. If you visit the cemeteries where my mother grew up, you will see that the men and women typically died in their 40's and 50's. My father died of natural causes just short of his 72nd birthday. My mother died at the age of 54 from cancer of the common bile duct, which had spread to her liver. It is a fact that certain chemicals cause cancer.

Cancer is Not Contagious

One belief that does need to be dispelled is that cancer is contagious. You can not become infected with cancer from someone else. Even if you drink out of the same glass as a cancer patient, you can not contract the disease. There is a similar myth circulating in some circles. This is that viruses cause cancer. There is no real evidence to support this belief. Because the Epstein-Barr virus became linked to leukemia in cats, some researchers embrace the notion that viruses can cause cancer. I can offer a different and more plausible explanation. It is more likely that cancer either already developed in the cat, which would make the cat more susceptible to the virus, or that the viral infection made the cat more susceptible to the development of cancer. An examination of the current literature will support either explanation.

Microorganisms other than viruses, such as yeast and bacteria, do not cause cancer either. Our bodies have too many defense mechanisms designed to destroy microorganisms both in the blood and inside the cell. Microorganisms have their own genetic material, it can not be substituted into ours. When the microorganisms are destroyed by enzymes, their genetic material is destroyed as well. In addition, our bodies do not have the enzymes required for the incorporation of genes from microorganisms. From a biochemical perspective, the enzymes in our bodies can not recognize the genes from microorganisms such that they can be expressed. There also is no evidence that any microorganisms contain genetic material that can initiate the formation of cancer. You do not have to be worried about getting cancer from any infectious agent, including viruses. You can not get cancer from a toilet seat.

Back-to-Nature

One cancer myth had its roots in the back-to-nature movement of the late 1960's. This movement included a groundswell of people devoted to cleaning up the environment. Part of their ideology involved using natural insectides to grow "organic" crops. The philosophy of growing organic was to let nature takes its course with a little boost from man. A number of researchers adopted this back to nature "organic" philosophy and as a consequence, two fad theories, basal cell and oncogene, sprung up in cancer. At the time, it was quick and easy for "organic" scientists to make a buck and to become recognized as innovative cancer researchers.

The Basal Cell Myth

One of these theories decreed, with no substantive data, that our bodies make cancerous cells all of the time; the formation of cancer is a natural process. The cells from which the cancers formed were called basal cells because they were suppose to have the basic structures of cells before special structures were put into place to form the various cell types. The corollary to this myth was that we are not all destined to develop cancer because our immune systems produce natural killer cells that continually seek and destroy cancer cells. Life threatening cancers are, therefore, due to a glitch in the immune system.

There is no evidence to support either contention. Both are based upon simplistic philosophy, not fact. This kind of thinking is fine for the vegetable garden, but it has no place in investigative science.

I have seen photographs taken through the electron microscope that allegedly show a white blood cell attacking a cancer cell as it enters a blood vessel. These pictures are still being used today as "proof" that the immune system seeks out and destroys cancer cells. From their faith in these pictures, immunologists reasoned that it is a simple procedure to cure cancer. All they have to do is find out how to augment the immune system to produce more white blood cells to fight the cancer. Billions of dollars have been spent to fruitlessly pursue this theory. Based on this basal cell theory of cancer, a number of products have been promoted as anti-cancer agents. Among these agents are the interleukins, the interferons, and monclonal antibodies.

This basal cell theory was founded upon spurious evidence at best. It was presented as a last ditch effort by cancer researchers at that time to save their projects. This theory came out at about the time Nixon declared war on cancer in the early 1970's. After about five years of research and billions of dollars, the public wanted some progress in any facet of the disease. The best information the researchers could give us at the time was a growing list of carcinogens, without being able to explain why those agents caused cancer. The public became impatient. It appeared that the experts did not know what they were doing. The news media announced, just in time, that researchers had found a promising breakthrough in cancer. That breakthrough was the basal cell carcinoma theory. This theory was thrown together to keep the scam in operation. For several years it successfully quieted the public.

After this theory was presented, the researchers went about obtaining data to support it. Instead of developing a theory based upon the data, they developed a theory and contrived supporting evidence. This just fueled the fraud for which we are paying dearly every day.

I saw the photographs of the white blood cell allegedly attacking the cancer cell. In all of these pictures, there was always one striking feature about the cancer cell.

This feature was present regardless of which type of cancer was being pursued by the white blood cell. All of the alleged cancer cells were spheres.

This observation may not seem important to you. The importance of it also escaped just about every other cancer researcher. When cancer cells grow in tissue culture, they are relatively flat and cling to the bottom of the flask by what appear to be tentacles. Some cell lines float in the medium in sphere-like colonies, but the individual cells are not spheres. Dead cancer cells in tissue culture are cells that float in the medium as single spheres; they are not attached to other cells. The researchers took pictures of the white blood cell destroying a dead cell, not a cancer cell. This is the mechanism our bodies use to get rid of dead cells in general.

When you have an operation and are then sewn back up, a number of cells inside your body die as a result of the operation. Those dead cells are removed by the white blood cells. If the spherical cell was a cancer cell as claimed, then it was a dead cell. Another feature that went unnoticed by other researchers is that at the magnification used, it is impossible to tell if that cell was in fact a cancer cell or a normal cell. No one can even determine which organ that cell came from.

There are other problems with the basal cell carcinoma theory. Our bodies are genetically programmed for the maintenance of life. When we get cut, the wound heals. When chemicals and microorganisms invade our bodies, we have elaborate defense systems to destroy them. But cancer takes life. What those researchers are telling us is that our bodies are programmed to self destruct more quickly than they should. This thinking runs contrary to common sense and to experience.

Another problem with this theory is in its implication that our bodies can attack themselves. If the white blood cells are always finding cancer cells in our organs, then it is possible that the same white blood cells can attack the normal cells of our organs. This means that our bodies have the capability to self-destruct. Some immunologists will argue that normal cells have special chemicals on their outer membrane that act as recognition markers to prevent attack by the white cells. Cancer cells lack these recognition chemicals, so they are attacked by the immune system. As evidence for this theory, the immunologists bring out pictures showing white blood cells surrounding tumors. The white blood cells may in fact be at the tumor, but they are not necessarily attacking live cancer cells. An alternate explanation is based on the fact that cancer cells grow rapidly. They can multiply so fast that they can outgrow the blood vessels that provide nutrients for them. Cancer cells consequently die at a faster rate than normal cells. The white blood cells in the photographs could be simply removing these dead cells. The most convincing evidence against the basal cell carcinoma myth may be that there has never been any consistent cure of a cancer patient using agents isolated from the immune system.

The Oncogene Fraud

The oncogene theory, the other back-to-nature offshoot, states that cancer forms because there are genes in each cell that code for a specific cancer. You get lung cancer, for example, because there is a gene that initiates the process of tumor formation in the lung. In the scientific literature the genes are represented by three letter symbols such as *mer* or *ras*. The three letters come from the type of cancer each gene codes for. Billions of dollars are spent each year to find new oncogenes. For each oncogene search, tons of radioactive waste are generated. Remember, a Nobel Prize has already been awarded to the two molecular biologists who laid claim to this theory.

The oncogene theory began in the early 1980's when the public again became impatient with the lack of real progress in cancer. All we kept hearing was the ever lengthening list of activities and chemicals that caused cancer. None of this provided a clue as to what caused cancer or how it could possibly be cured. The credibility gap was increasing between cancer researchers and the public who wanted to know how their money was being spent. In an act of desperation to keep their respective projects afloat, molecular biologists invented the oncogene theory. After the details of the theory were developed, they set out on a course to manufacture data to support it. I have already described how easy and routine it is for molecular biologists to falsify their data to "prove" they found a gene. This is exactly the process they employed. Molecular biologists may find actual alterations in the genetic code, but these can be explained without resorting to fabrications, as you will learn.

A curious tactic of oncogene proponents is to present data that refutes their theory and then claim that it supports it. A summary of work with human breast cancer cell lines by a physician who believed in the oncogene theory is an excellent example of this absurdity. This work was presented at a cancer convention I attended in Atlanta.

The presentation was particularly interesting to me because her work in human breast cancer cell lines paralleled what I was doing in human lung cancer cell lines. Naturally I was curious to find out if she obtained results similar to mine, because my theory predicts that she would have to. She did.

She had isolated two breast cancer cell lines. One grew only in standard tissue culture, the other when estradiol was added to the medium. As she increased the amount of estradiol in the medium, the rate of both cell lines increased. I had found a similar pattern with the lung cancer cell lines but with a different chemical messenger, serotonin. Wishing to be scientifically fashionable, she reasoned that one cell line was estradiol dependent because it lacked a necessary oncogene. This was already a contradiction in the oncogene theory. If the cell line lacked the required oncogene, then how did it become cancerous?

In an effort to transform the estradiol dependent cell line into the non-estradiol dependent cell line, she invited a molecular biologist friend to put the right oncogene into the estradiol dependent cell line to promote this transformation. The physician also had normal human breast cell lines growing in culture. The molecular biologist put his alleged oncogene into that third cell line as well. Now both the estradiol dependent and normal breast cell lines had an infusion of an oncogene which was suppose to transform them into the estradiol independent cell line. Guess what happened? *Nothing.* There was no transformation of either cell line. Neither the normal cell line nor the estradiol dependent cell line transformed into the carcinogenic cell line. Nevertheless, the physician still went on to report her belief in oncogenes when her own data clearly demonstrated the contrary. Here we have an experiment, designed to prove the existence of oncogenes, that backfired, but still the researcher mindlessly maintained the hoax.

There are many similar experiments reported in the literature that also definitively argue against the existence of oncogenes, while the authors somehow twist the results into a support for this theory. For example, it is commonly known among researchers that when an animal is dosed with a chemical carcinogen, specific organs are targeted to induce the formation of tumors. Tumors do not form indiscriminately throughout the animal. A reputable molecular biologist described how she dosed mice with a nitrosamine known to be a potent carcinogen. She then measured the expression of the oncogene, or so she claimed. She found that the oncogene she was interested in was expressed in all of the organs of the mice. However, tumors formed only in the pancreas and liver. What happened to the other organs in which the oncogene was expressed? Your guess is as good as mine, and better than the molecular biologist's.

Simple logic would seem to dictate that tumors had to form in all of the organs, because according to the oncogene theory, cancer forms whenever oncogenes are expressed. This researcher reported that the oncogene was expressed. Tumors should have been seen in every organ. The reason that this did not happen is that oncogenes do not exist. We are not genetically programmed for cancer. The very pieces of evidence that the molecular biologists offer to support the existence of oncogenes is always inconsistent with the very theory they believe in.

For the sake of discussion, let us allow the molecular biologists their oncogenes. To date, they still can not tell us what turns oncogenes on. They might suggest that chemical carcinogens somehow perform this function. The question remains, "How?" They have no answer. All molecular biologists have succeeded in doing is to place another unnecessary step into the proposed initiation process of tumor formation. Because they can not explain how oncogenes are turned on, we are back to square one. The question remains, "What causes cancer?" Not the mysterious expression of oncogenes decreed by them.

The Heredity Hoax

Another myth circulating in the popular press is that cancer is genetically inhereted. Again, we have a fad belief with no real data to support it. Many of you may be thinking, "How can he say this? My grandmother had breast cancer; my mother had breast cancer; the physicians tell me I will get breast cancer because it is in my genes." Physicians have less understanding of molecular biology than molecular biologists, most of whom have embarassingly little. The scientists propagating this myth of genentic inheritence are inept and incompetent, but they continue to con you.

Think about it. Soon after you were born, did your parents pack you in a shipping crate for exportation to the other side of the country? Or did your parents raise you in basically the same environment that they grew up in? For most people, the latter is true. Consequently, you were exposed to the same carcinogens as your parents. It is true that some of you may be more genetically predisposed toward a reaction to a given chemical carcinogen. This is not a new or unusual phenomenon. Some people have an allergic reaction to aspirin. Some people are allergic to bee stings or to penicillin. A greater sensitivity to a specific chemical is a well documented observation. A history of a particular cancer in some families is due to environmental exposure coupled with greater sensitivity to a carcinogen. The only exception to cancer susceptibility is radiation, such as from ultraviolet light and radioactive compounds. Anyone can be overexposed to radiation and get cancer. Your genes have nothing to do with this.

In short, if there is no exposure to a carcinogen, you will not develop cancer. This statement is true regardless of your genetic background. The data which was gathered to support the theory that cancer is genetically inherited was based upon poor sociological studies. Those studies were so ineptly conducted that none of them would be acceptable for even an undergraduate research project today.

Cancer can not be transmitted like an infectious disease. It is not caused by a virus or any other microorganism. You do not continuously form cancer cells which your immune system attacks. None of us is sitting on a time bomb such that a certain number will definitely develop cancers. There are no genes that code for the formation of cancer. No one is genetically programmed to get cancer. These theories were created as desperate attempts by some cancer researchers to continue their projects and to feed their egos with a cheap shot at international recognition. No matter how vehemently they protest that their theories are valid, the real evidence, and scientific principles, clearly show that their theories are false.

Cancer Prevention Fads

The myths explaining cancer's cause are routinely coupled with fads in its prevention. The fact is that *the only way to prevent cancer is to avoid exposure to the causative agents*. These causative agents are certain chemicals like the ones present in tobacco products and vehicle exhaust emissions and radiation. Eating the "right" types of food not will prevent cancer. All of the fad cancer preventive measures prescribed to you by the "experts" have no effect against the formation of cancer.

High fiber in the diet does not prevent the formation of colon cancer. I conducted an experiment with human lung cancer cell lines which included one of the chemicals, sinigrin, found in vegetables such as cabbage, broccoli and cauliflower that is suppose to inhibit the metabolism of carcinogens, and, thereby prevent the formation of cancer. No inhibition of carcinogen metabolism was observed. The notion presented by the so-called experts in cancer and nutrition arose from the fact that this chemical is present in a relatively large quantity in those vegetables. This followed their observation that central Europeans who eat large amounts of those vegetables have a relatively low incidence of colon cancer. Misusing statistical analysis, they concluded that some chemical in the vegetables prevented the metabolism of carcinogens and therefore the formation of colon cancer. The results of my experiments refuted this claim. A diet consisting of fruits and vegetables is healthful for you, but it will not prevent the formation of cancer. More recently, the NCI decreed that eating broccoli reduces the incidence of breast cancer in women. This is another politically expedient move by cancer researchers to obtain favor from women to support their bogus research programs. I already discussed how easy it is to manufacture statistically correct data and to even falsify data. Simply put, **there is no diet that can cure or prevent the formation of cancer**. Please, entertain no false hopes on this matter.

The focus of attention on ways to prevent cancer is the result of one of the directors at the NIH who stupidly concluded that no one will ever discover the mechanism of cancer and, therefore, that no one will ever be able to cure this disease. In an unofficial directive, so I was told, he had researchers concentrate on finding ways to prevent the disease. Here we have yet another unsubstantiated theory that is nothing more than a decree by an incompetent, but powerful, alleged expert. Again, evidence was manufactured to fit theory. The bottom line is that if you want to avoid getting cancer, avoid exposure to known carcinogens. If you have a family history of a specific cancer, then find out what carcinogens those family members may have been exposed to, and avoid them. People with a family history of skin cancer obviously have a predisposition to the damaging effects of ultraviolet rays. Ultraviolet light is a genuine carcinogen that can affect anyone excessively exposed.

Bad Theories Beget Bad Treatments

There is a rhyme and reason to cancer, but the bogus theories being circulated hinder the process of obtaining the truth about this disease. In the next two chapters you will learn how and why all current cancer treatments to date must necessarily fail. As you may be able to guess by now, the theories upon which those treatments are based also were fabricated to be fashionable in order to protect a given researcher's project, and so that his ego could be fed.

CHAPTER 16

THERE IS NO CURE FOR CANCER

There is no cure for cancer, yet cancer can be cured. Some of you may be thinking that you were right all along, the government has been concealing this fact for decades in order to allow physicians and hospitals to make billions of dollars. Others may have given up hope that this disease can truly be conquered. Before you draw any conclusions, let us examine my initial statement more closely.

There Is No Cure, But...

The word "no" means none. Cure is presented as a singular noun. Therefore, there is no single drug or treatment that can be used against all types of cancers. While the word "cancer" is in the singular form, it is not used in the context familiar to the public and researchers. As a general disease state, cancer can be cured. However, for each and every type of cancer there must be a different specifically designed drug. The correct strategy for curing cancer is to think in terms of *cures* for *cancers*.

Curing cancer is a much more difficult problem than anyone has ever envisioned. The situation is complex because an anti-cancer drug has to target a specific cell type, not just the organ in which the tumor is located. To illustrate the complexity of the problem arising from the different cell types of cancer, Table 16-1 is given as a summary of data three colleagues and I published.

In this article, we presented data on the differential rates of metabolism of the carcinogen diethylnitrosamine (DEN) by several human lung cancer cell lines. The human lung is comprised of about 20 different cell types working in unison to enable us to breathe. Each cell type performs a different function during this process. DEN has been strongly linked as one of the major chemical carcinogens initiating the formation of lung cancer in persons who smoke tobacco products. Of the four cell types, or lines, presented in Table 16-1, the NCI-H727 is the cell type responsible for forming tumors in people who smoke. It was logical for us, therefore, to examine how the different cell lines metabolize DEN. To monitor this metabolism, we used a ^{14}C tag, which is a radioactively labelled carbon atom at a specific location in the DEN molecule. We measured the rate at which this tag was given off, expressed as $^{14}CO_2$ pMole/10 Minutes/mg protein in the table. This technical term means that the higher the number, the faster the metabolism. In general, different rates of metabolism indicate how specific cell types respond to dosing with a chemical. Researchers have very precise methods for measuring this response.

The rates of DEN metabolism in Table 16-1 vary greatly from none detected in the NCI-H128 cell line to a very rapid rate in the NCI-H727 cell line. The rate in the NCI-H727 cell line is about 200 times faster than the rate in rat liver cells, the standard that biochemists insist on using. It is easy to underdstand why relying solely on data from rat liver cells to examine the metabolism of chemicals can allow valuable data to be missed. The rates in

Table 16-1 clearly demonstrate that not all human cell types are the same from a biochemical perspective. Remember, these are cancer cells, which are not as biochemically intact as their normal cell counterparts. Normal cells may have even greater variations in the response to the metabolism of DEN and other chemicals, including drugs.

Table 16-1. Metabolism of ^{14}C-DEN by Human Lung Cancer Cell Lines

Cell Type	Morphology	$^{14}CO_2$ pMoles/ 10 Minutes/ Mg Protein
NCI-H128	Small-cell cancer	None Detected
NCI-H358	Adenocarcinoma: alveolar type II	466 ± 18.4
NCI-H322	Adenocarcinoma: Clara cell	829 ± 21.1
NCI-H727	Carcinoid: Neuroendocrine cell	$28,925 \pm 56.7$

Reprinted from the Proceedings of IXth International Symposium on N-Nitroso Compounds held in Baden, Austria during 1-5 September, 1986, sponsored by the World Health Organization of the International Agency for Research on Cancer (IARC): Cell Type-Specific Differences on Metabolic Activation of N-Nitrosodiethylamine by Human Lung Cancer Cell Lines. H.M. Schuller, M. Falzon, A.F. Gazdar, and T.J. Hegedus, p. 138-140.

For the sake of discussion, let us pretend that DEN is an anti-cancer drug instead of a carcinogen. Four patients come into the hospital coughing and complaining of difficulty in breathing. Standard biopsy procedures using the light microscope would indicate that all four patients have small cell cancer of the lung. DEN is prescribed by the physician to treat the cancer. How do you think the patients will respond to the treatment?

If one patient has lung tumors derived from NCI-H128 cells, then no DEN will enter into the tumor cells and the patient will die from lung cancer. If one patient has tumors derived from NCI-H727 cells, then there is a good chance that the disease will be cured because a large amount of the DEN "drug" will enter the tumor cells. If the other two patients who have tumors derived from either the NCI-H358 or NCI-H322 cell types, each of these patients will have a shrinkage of the tumor, but they will not be cured of the cancer. This pattern of response is typical of the clinical situation as it exists today. A few patients respond to therapy, but the overwhelming majority receive only temporary benefit from treatment; while almost all patients have no response. Identification of the cell type is crucial to knowing how to treat this disease.

As the diagnosis is made according to current procedures, it is not possible for the pathologist to determine the cell type of the tumor. In fact, most physicians and cancer researchers do not believe that determining the cell type is important. To them, all cancers are the same. Clearly, the evidence shows this thinking is false and outdated. Pathologists make their diagnosis using the light microscope, which is similar to the kind you see in the movies and television shows. The upper limit of magnification is about 2,000 times and the instruments costs from $3,000-$10,000. In order to determine the cell type, an electron microscope has to used to view key features, called ultrastuctures, inside the cell. Differences in the ultrastructures determine the cell type, but they can only be viewed at magnifications of at least 15,000 times. Electron microscopes can magnify up to about 200,000 times.

Electron microscopes are not used in the diagnosis of cancer for two reasons: cost and lack of expertise. An electron microscope costs between $100,000 and $150,000. While it takes about a day to prepare samples for viewing with a light microscope, about 3 days are needed to process samples for the electron microscope. Specially trained technicians have to be hired to prepare these samples. Also, very few pathologists are skilled in diagnosing specimens viewed with the electron microscope. So the easier and cheaper light microscope is used. In reality, a biopsy can not confirm that a patient has cancer. This procedure can only determine that a rapid growth of cells is present in the organ. This rapid growth can be either cancer or hyperplasia (excessive proliferation, i.e., growth, of normal cells) in that organ. Even if the results of the biopsy indicate that the patient has cancer, the cell type of cancer can not be determined. This lack of information can make the difference between a patient living and dying.

People have asked me why some individuals are spontaneously cured of their cancer. One major reason is that the patient was diagnosed by the biopsy procedure as having cancer actually had a hyperplasia of *normal* cells in that organ. Hyperplasias are not cancers; they are not life threatening conditions. This type of growth is reversible and disappears spontaneously. A misdiagnosed hyperplasia will be interpreted by physicians as a "cure" for a cancer the patient never had in the first place.

The isolation of the NCI-H727 cell line resulted from a misdiagnosis of a lung tissue sample using the light microscope. Standard biopsy procedures indicated that the tumor was a malignant form of small-cell cancer. Based upon this diagnosis, the growth was removed and this cell line established. When diagnosed using the electron microscope, it became evident that it was a benign (slow growing) tumor. Fortunately, no harm came to the patient due to removal of the tumor. This is a real example of how patients can be misdiagnosed if the pathologist does not use the appropriate equipment to get the job done.

All cell types, including cancers respond differently to alterations in the structures of chemicals. Generally, as the structures of the chemicals become more complex, the responses by cell types become more specific. This concept can be illustrated by the same set of human lung cancer cell lines that were used in the DEN experiment. The NCI-H358, NCI-H322, and NCI-H727 cell lines were dosed with the exact same concentration of nicotine, and then the growth rate was monitored. Nicotine has a more complex structure than DEN. Therefore, different responses by the cell lines, expressed as different growth rates, were expected. To monitor this growth rate, the cells were counted every 3 days using the technique described in the home grown cancer research. The results of these experiments are summarized in Table 16-2. A portion of this work was published in the journal Toxicology In Vitro, in 1989.

For each of the cell lines, we monitored the normal growth rate to serve as the control group. If you look in the table, you will notice that the NCI-H358 cell grows the fastest, and the NCI-H727 grows the slowest. These differences in growth rate partially explain why some cancer patients die quicker than other patients. Additional important information is contained in this table. Look at the variable response to nicotine by each of the cell lines. Again, imagine nicotine as being an anti-cancer drug.

The effect of nicotine on the two cell lines NCI-H358 and NCI-H322 is minimal. After 3 days, the cell counts are about the same in the control and nicotine treated groups. At days 6 and 9, the cell counts vary slightly from the control to the treated goups, but this variation is not substantial. Statistical analysis of this data would have indicated a significant response between control and treated groups of both cell lines. These differences can readily be explained by the unavoidable experimental error associated with this kind of test system. This data also illustrates the ease by which statistical analysis can be misused; or to put it another way, how a researcher can lie with statistics. Other cancer researchers would conclude that these are interesting results that warrant additional work to evaluate the effectiveness of nicotine as an anti-cancer drug. You would pay for this trivial pursuit.

Of the three human lung cancer cell lines, only the NCI-H727 had a dramatic response to dosing by nicotine. The cell count doubled after 3 days, quadrupled after 6 days, and plateaued to about 2.5 times the normal growth rate after 9 days. This leveling off of the growth rate is a common observation. Even without using statistical analysis, it is evident that nicotine has a tremendous effect on stimulating the growth rate of the NCI-H727 cell line. We did not need statistics to prove our point.

Smokers should heed the warning implicit in this data. The cell type responsible for about 70% of lung tumors that develop in smokers come from this NCI-H727 cell type. Nicotine, which is found in all tobacco products, not only initiates the formation of lung cancer derived from neuroendocrine cells, but it also accelerates the growth rate of the tumor if that person continues to smoke. Some of you may have read in the newspaper that researchers believe there may be a link between cigarette smoking and an increase in the size of the tumor. Two years before this article appeared, we not only demonstrated this link, but also quantitated how fast the tumor growth rate can be accelerated.

Table 16-2. Effects on Growth in Human Lung Cancer Cell Lines Dosed With 1 μM Nicotine

Cell Line	Treatment	Number of Viable Cells x 10^4 at Days:		
		3	6	9
NCI-H358	Control	5.5 ± 0.2	13.5 ± 0.5	50.8 ± 0.3
	Nicotine	5.6 ± 0.8	14.5 ± 1.0	57.3 ± 1.0
NCI-H322	Control	5.3 ± 0.1	10.2 ± 0.3	43.5 ± 0.7
	Nicotine	4.4 ± 0.3	11.3 ± 0.2	32.3 ± 0.8
NCI-H727	Control	5.5 ± 0.2	8.8 ± 0.3	38.5 ± 1.8
	Nicotine	11.0 ± 1.0	36.3 ± 1.3	97.9 ± 2.9

Again, if we pretend that nicotine is an anti-cancer drug, tumors derived only from the NCI-H727 cell type would have responded. Only patients with this type of cancer would be cured. No response would have been observed in patients with the other two types of cancer. They would die from lung cancer. Based upon these two sets of data, from the experiments on DEN metabolism and cell growth kinetics after dosing with nicotine, it is clear that the response by cell types to the same chemical is measurably different. This differential response is the primary reason that there is no cure for cancer.

There Are Cures

Nevertheless, cures for cancers can be developed by talented medicinal chemists once their fascist bosses are removed from power in the medical research community. Physicians will never cure cancer because their main method of research is the misuse of statistics, and most importantly, they have a limited understanding of chemistry. Molecular biologists can never observe the difference in the biochemistry of cells because they have narrowed their focus of research exclusively to mindlessly isolating DNA fragments and, to a lesser extent, RNA molecules. And, they have virtually no understanding of chemistry despite the fact they stumbled across the spelling of molecular, which implies knowing chemistry. I wonder if they know how to spell pharmacology. They completely destroy an organ from which they extract DNA in a futile search for non-existent oncogenes. Due to this extraction procedure, the necessary kind of information like that presented in the two tables above is lost. Chemists regard these biochemical differences in cell type merely as an intellectually curious set of observations, but inconsequential for their "rational" development of anti-cancer drugs. I consider these differences as vital information required for my rational development of anti-cancer drugs.

In the future, when you read or hear a story that some researcher has a cure for breast, colon, or lung cancer, you now know better than to believe this report. This alleged cure is pure hype to play upon the emotions of hope and kindness to con you out of money. Their game is slick, but now you have information which you can use to counter their moves. Organs are comprised of several cell types. The specific cancerous cell type must be determined by electron microscopy and verified by pharmacological tests to make a positive identification before the patient can be given the appropriate drug. Unfortunately, none of these drugs exists at this time.

There is no "magic bullet" that will cure all cancers. Various cancer treatments are reviewed in the next two chapters with the reasons each must fail based upon proven scientific principles and theories. Cancer can be cured, but each type of cancer must be attacked differently. Now you can understand why I stated that curing cancer is a much more complex problem than anyone ever imagined. Targeting organs is a useless and impossible strategy. Cell types must be targeted to cure cancers.

CHAPTER 17

WHY ANTI-CANCER AGENTS FAIL

Selective toxicity is the main goal in cancer treatment. The ideal anti-cancer drug must be able to kill the cancer cell type while exerting no effect on the normal cell counterpart. Almost all anti-cancer agents used to date are highly toxic to cells. The problem is that they are incapable of killing only cancer cells. These agents generally kill all life forms. As a result of this indiscriminate toxicity, cancer patients experience a large array of adverse side effects including nausea, vomiting, blood disorders (dysplasias), weight loss, hair loss, neurotoxicity, coma, other cancers and even death. Each of the treatments used against cancer must fail. The reasons for failure are based upon well established principles and theories in chemistry, biochemistry and pharmacology. The mechanism of action is an explanation of how an agent exerts its effects. This will be discussed for each treatment. In efforts to improve selectivity, researchers have examined a variety of agents including interleukins, interferons, monoclonal antibodies, chemotherapeutic drugs, gene therapy and electronic toys such as lasers. Other researchers have experimented with combinations of anti-cancer agents and dosage regimens. While some of these researchers may have the best of intentions regarding help for cancer patients, none of these strategies can cure cancer.

The fundamental principle guiding the utilization and design of all current anti-cancer treatments is to take advantage of the fact that cancer cells grow faster than their normal cell counterparts. Due to this increased growth rate, the cancer cells are suppose to take up the anti-cancer agent at a greater rate than the normal cell. In essence, this approach to curing cancers boils down to a "footrace" between killing all of the cancer cells before significant, irreparable damage is done to the organs of the patient. The gamble is that the cancer will be cured before the patient dies from the treatment. I wish I could tell you that the strategy of cancer researchers is more sophisticated than this, but it is not.

As I discussed in the previous chapter, no single agent can cure all cancers. The misguided focus of researchers on one agent to cure all cancers will become evident as each treatment is described in detail. Most of these agents were tried first on cancer cell types grown in tissue culture before they were administered to patients. When you read the suffix "oma" in the subsequent discussion, it refers to a cancer of that cell type. Cell types will be presented as some combination of letters and numbers much like the make and model of a car. When you read this designation, just think of them as different makes and models of cancer cell lines. Let us examine how current anti-cancer agents exert their toxic actions, why those actions are not selective for cancers, and why these agents fail as cures.

Interleukins

Interleukins (IL) are large proteins manufactured by some of the white blood cell types of the immune system. Two major types of IL's have been isolated and designated as IL-1 and IL-2. In addition, each of these two types are further subdivided into alpha and beta forms. Our immune systems produce IL-1alpha, IL-1beta, IL-2alpha and IL-2beta. Biological functions of IL-1 are the induction of IL-2, expression of IL-2 receptors in activated T-lymphocytes (a type of white blood cell in the immune system), induction of B-cell proliferation (another type of white blood cell), increasing natural killer cell activity (a third type of white blood cell), regulating antibody dependent cytotoxicity (i.e. germs are recognized as foreign to the body and then killed by the immune system), and allegedly the killing of tumor cells.

All of the biological functions of IL-1 can be summarized as simply the stimulation of various parts of the immune system in response to invasion by microorganisms. The rationale for stimulating the immune system to cure cancer is that some researchers have proclaimed that cancer cells are mutations of normal cells, and, therefore, foreign to the body. Consequently, if our immune systems are "trained" to recognize cancer cells, those cells will be destroyed naturally. By isolating agents within our immune system responsible for its regulation, researchers claim they can improve the targeting of white blood cells to attack cancer cells. There is no real scientific data to support this proclamation that cancer cells are mutations. If cancer cells are mutations as decreed, then why doesn't the immune system consistently attack and destroy all cancer cells? The real explanation is that cancer cells are not mutations. They are hastily constructed cells. Therefore, "activating" the immune system against cancer is a useless strategy at best, lethal in the extreme, because normal cells in that organ also will be attacked.

IL-1 was isolated from a leukemia cell line THP-1 in the alpha and beta forms and from a mouse osteoblastic (bone cancer) cell MC3T3-E1 in the beta form. The IL-1beta isolated from the mouse bone cancer cell line was a recombinant form derived from human IL-1 and mouse IL-1. The researchers claimed that IL-1beta inhibited DNA synthesis and cell growth of the same mouse cancer cell line from which it was isolated.

Another group of researchers tested IL-1beta against six different types of cancer cell lines obtained from humans. These included three human breast cancer cell lines, MCF-7, MDA-MB-415 and T47D, a human lung cancer cell line, CALU-1, a human colon cancer cell line, SW-49, and a human embryonic lung fibroblast cell line, HEL. Of these six human cancer cell lines, IL-1beta partially inhibited the growth rate of one breast cell line, MDA-MB-415, and to a substantially lesser extent the growth rate of the other two breast cell lines, MCF-7 and T47D. No inhibition of the growth rate was observed in either of the two human lung cancer cell lines nor in the colon cancer cell line.

These two studies are not an exhaustive review of the work done with IL's. They are, nevertheless, representative of this type of cancer research. In all of the articles I read concerning IL research, the authors claimed that their IL's were a successful treatment for all cancers in that their results offered promising new leads, despite the fact that most cancer cells types were not killed in the controlled environment of tissue culture. If the cancer cells are not killed by IL's in the flask, they certainly will not be killed in your body.

Because the IL's noted above were isolated from a leukemia cell line and a mouse bone cancer cell line, a logical question is why the IL's didn't kill the cancer cells from which they were isolated. An explanation could be that the IL's are in an inactive state when stored in the cancer cells. Once the IL is released, it could exert its lethal effects. There is some precedence for this explanation. The lysosomes described in the chapter on the cell act in this manner. The main difference between the actions of lysosomes and IL's is that lysosomes destroy all life forms whereas IL's obviously do not. The other plausible explanation is that IL's are not anti-cancer agents, even though they are manufactured by the immune system in response to invasion by microorganisms. In the study that tested IL-1beta against six human cancer cell types, only one cell line was even somewhat affected. This agent was virtually useless against the other five cancer cell lines.

Interleukins are toxic proteins. As evidence of this toxicity, another study reported that IL-1 stimulates the following liver proteins: C-reactive protein, serum amyloid, fibrinogen, $alpha_1$-acid glycoprotein and metallothioneins. The stimulation of these liver proteins indicates liver toxicity or damage.

Due to the large size of IL's, these proteins never enter the cancer cells. IL's destroy microorganisms because the white blood cells engulf and absorb the microbe. Once the invader is inside of the white blood cell, then the IL's are released to destroy the microorganism. Cells not of the immune system do not have this ability to engulf large objects.

One major adverse side effect with IL treatment is that the growth rate of both normal and cancerous T-helper cells of the immune system is stimulated. This effect means that if a patient is treated for lung cancer with an IL, that patient might develop leukemia. Meanwhile the patient is never cured of the lung cancer. Another dangerous side effect resulting from IL-1beta treatment is the stimulation of bone resorption. Translation of this effect is that your bones become thinner and more brittle, leading to osteoporosis.

Researchers claim that IL's inhibit the synthesis of DNA. As I discussed in previous chapters, there is no significant chemical difference between DNA found in normal cells from that found in cancer cells. If the inhibition of DNA synthesis leads to cellular death, then both normal and cancer cells will die.

Some researchers argue that with more money to fund further work they will be able to genetically engineer purer forms of IL that will reduce the incidence of side effects by targeting tumor cells. This claim is not true. The logical end result of "training" the immune system to attack an organ that has cancer is to also alter the immune system to attack normal cells of that cell type throughout the body with corresponding adverse side effects. For example, neuroendocrine cells are found in the pancreas, lung, thyroid, and adrenal glands. Neuroendocrine cells manufacture an important chemical messenger called serotonin required to help keep the biochemistry of our bodies functioning properly. If the immune system is altered to attack cancerous neuroendocrine cells of the lung, then the normal cells also will be destroyed in the lung and in each organ having this cell type. This destruction will set into motion a cascade of events leading to the development of adverse side effects. The array of side effects may be different with IL therapy, but the severity of the effects will not change. The price tag for genetically engineering IL's could be as high as $150 million just to locate an individual IL gene.

Interleukins have no real utility as anti-cancer agents. Experimental data shows that these proteins are not selective toward any form of cancer, nor do they kill most types of cancer. Adverse side effects include liver damage, bone resorption, leukemia and the possibility of non-specific destruction of healthy organs, even without tumors present. The fact that IL's are isolated from cancer cells is strong evidence to support the conclusion that they are not anti-cancer agents. When one considers the costs of producing IL's, their limited toxicity toward all types of cancer cells and the terrible adverse side effects, it is obvious that this form of cancer treatment should not be pursued.

Interferons

Interferons (INF) are another group of protein hormones isolated from the white blood cells of the immune system. There are three types of IFN's labelled as alpha, beta and gamma. The alpha and beta forms of the IFN's should not be confused with the alpha and beta forms of the interleukins. While all three forms of the IFN's stimulate the immune system, alpha-IFN is the most potent. The actions of IFN's have been linked to antiviral, antimitogenic and immunomodulatory activities. Antiviral is self-explanatory in that it means IFN's attack and destroy viruses. Mitogenic refers to the process of cell division; antimitogenic means that IFN's inhibit the division of cells. Immunomodulatory simply means that IFN's can either stimulate or inhibit the formation of and biochemical processes in white blood cells.

The exact mechanism of action by which IFN's exert their effects is uncertain. Researchers have found that IFN's inhibit the translation of RNA to code for protein synthesis. Also, the two enzymes protein kinase and phosphodiesterase, required for protein synthesis, are inhibited. In one study, which clinically tested IFN's against a variety of cancers, the researchers concluded that the mechanism of action for tumor regression could not be explained. Most cancers had no response to IFN treatment.

Interferons alpha and beta were tested clinically against the following cancers: hairy cell leukemia, lymphoma, melanoma (a malignant, pigmented mole or tumor on the skin), Kaposi sarcoma (another type of skin cancer), renal sarcoma, myeloma (multiple tumors in the bone marrow), breast cancer, gastrointestinal cancer, pulmonary (lung) cancer, metastatic genitourinary cancer and papilloma (warts). Of these 12 types of cancer, the hairy cell leukemia had the best response, but a total remission or cure was not achieved. As postulated by the researchers, the mechanism for the limited success against this type of leukemia was that the IFN stimulated the activity of yet another group of white blood cells labelled natural killer (NK) cells. The lymphoma also had a very limited response to IFN treatment. The researchers explained this response as being due to the NK cells producing IL-2 and B-cell growth factor. The B-cell produces IL-2.

At this point your head may be spinning trying to keep track of the different types of white blood cells that make up our immune system. Do not be upset; it is confusing. All of this data can be simplified to the following understanding. There are two major groups of proteins produced by our immune system, interleukins and interferons. These two proteins are responsible for sending messages to other cells in the immune system to first make more proteins to fight infectious diseases and then to make more cells to fight these diseases. Cancer researchers want to take advantage of these biochemical processes to fight cancer. Most recently, researchers are trying alpha-IFN, wishing it might be more toxic and selective toward cancer cells. The fact that no one has been cured of cancer using this IFN, or any other, indicates that their wish will not be fulfilled.

Although the exact mechansm of action is not known, researchers claim that IFN's interfere with the transcription of RNA into proteins.Two enzymes, a protein kinase and a phosphodiesterase, are required to function properly for this transcription to occur. In all life forms, including microorganisms, plants and animals, this process of RNA transcription to synthesize proteins constantly takes place. Provided that the IFN enters the cell to inhibit the functioning of these two enzymes, RNA transcription will be inhibited in all life forms, including normal cells. This indiscriminate inhibition hardly meets the criterion for selectivity. A chemical that kills virtually all life forms is called a general protoplasmic poison. This capability of IFN's to kill all life forms is one of the mechanisms utilized by the immune system to defend our bodies from invading microorganisms. But, cancer forms within our bodies. Employing a system designed to fight invaders to attack parts of our bodies is illogical and doomed to failure, including an alleged vaccine against cancer. The overwhelming amount of experimental data from both tissue culture experiments and clinical trials on cancer patients confirms the inability of IFN's to cure cancer.

IFN's are harvested from the cells of the immune system including cancer cells. The question remains as to why IFN's do not kill the cancer cell types from which they were harvested. The strategy researchers are using with IL's and IFN's can be readily summarized. They harvest two toxic proteins from both normal and cancer cells of the immune system. These two toxins are then suppose to attack the same cancer cell types from which they were harvested with the expectation that only cancer cells will be killed. Somehow, the extraction of the IL's and

IFN's from the cancer cells are suppose to make these proteins selectively attack cancer cells. There is no scientifically rational explanation for how this transformation is to occur. The experimental data shows the folly of the researchers' logic. Like the IL's, IFN's are not selectively toxic to any type of cancer. In fact, there was no tumoricidal activity in most cases. The failure of IFN's to be effective against cancer can be easily explained according to proven scientific principles. In order for either IFN's or IL's to exert any toxic effect, the proteins must first enter into the cell. As described in the previous chapter on the cell, every cell wall is designed to keep out chemicals and microorganisms not required by the cell. As a general rule of thumb, the larger the size of the molecule, the more difficult it is for that chemical to penetrate the cell wall. Both IL's and IFN's are huge molecules. To give you an idea about the size differential between a molecule such as nicotine and IL's or IFN's, let us assume that nicotine is the size of a model airplane sitting on a shelf in your son's room. An IL or IFN would be at least the size of a Boeing 747 jetliner. If for no other reason than he can not bring the jetliner through the front door of your house, it would be difficult for your son to place a 747 commercial liner on a shelf in his bedroom. In an analogous fashion, cancer researchers want the gigantic IL's or IFN's, to enter an opening designed to accommodate very small molecules. A major reason there is no success against most cancers is that neither IL's nor IFN's can enter into the cell. However, cells of the immune system, even those that are cancerous, are designed to engulf huge objects like proteins and microorganisms. It is relatively easy for these cells to take in a large molecule like IL's and IFN's; therefore, there is some success against leukemia.

Entrance into the leukemia cell only partially explains the mechanism of action by IFN's. Enzymes catalyze chemical reactions to form new chemicals. There is an important feature about the products of these reactions. If an excess of product forms inside of the cell, the surplus of this product will cause the enzyme that made it to inhibit further production. This inhibition is temporary, but it will continue until the excess product leaves the cell. Biochemists call this process feedback inhibition. With interferon treatments, cell division is halted because excess IFN inhibits the transcription of RNA. At the same time, some of the cancer cells die due to the natural life cycle of the cell. This two stage process of cell division inhibition and natural cellular death leads to a decrease in the number of cancer cells. The patient is not cured of the cancer because the inhibition process is temporary and not lethal. As long as the patient receives IFN therapy, the rate of cancer cell growth is slowed but the cancer cells are never killed.

This is my explanation of the mechanism of action by IFN's. Well established principles of biochemistry, cell biology and pharmacology were applied in the development of this explanation. Proper application of these principles predicts that IFN's can not enter into cell types other than those of the immune system. The inhibition of cancer cell growth also is adequately explained. Had the researchers taken the time to think about the mechanism of how IFN's work before initiating expensive, ineffective treatments, then billions of dollars would have been saved and channeled to legitimate strategies for curing cancer.

Lack of selective toxicity is not the only limitation of IFN therapy. In clinical trials IFN's have produced mixed results in the modulation of natural killer (NK) cells: increase, decrease or no discernible pattern. Translated, these results mean that if a cancer patient is given IFN in an effort to stimulate the activity of the immune system against the cancer, no one can predict how the immune system will respond. The net result is that the patient may or may not receive benefit from this therapy. The patient has to spend tens of thousands of dollars on a crap shoot.

The list of adverse side effects due to IFN therapy is staggering: enhanced excitability of nerve cells, which can produce tremors; bradycardia, a potentially lethal slowing of the heart rate; disabling fatigue; central nervous system toxicity, exhibited as disordered cognition and memory (you can't think straight); somnolence (you can't sleep well); mood changes, like depression; fever; headache; myalgia (muscle pains); anorexia; nausea; vomiting; diarrhea; hypotension (low blood pressure); bone marrow depression (decrease in the number of blood cells manufactured), with acute granulocytopenia (abnormal reduction in the number of white blood cells) and thrombocytopenia (abnormal decrease in the number of platelets which are required for the blood to clot normally when you are cut); acute cardiac failure (you can die suddenly because your heart stopped beating); acute pulmonary toxicity (you suddenly have trouble breathing); renal failure (your kidneys stop working); and *unexplained death.* Central nervous system (your brain) adverse side effects include parethesias (abnormal sensation without an objective cause, such as numbness, prickling, tingling and heightened sensitivity), weakness, decreased attention span, short-term memory impairment, confusion, personality changes and coma.

All of the definitions for the medical terms in this book were taken from Taber's Cyclopedic Medical Dictionary, 12th Edition, copyright © 1973 by F. A. Davis Company, 1915 Arch St., Philadelphia, PA 19103. You can purchase a copy of this dictionary at any medical school bookstore for about $20.

Interferons are useless against most types of cancer. The minimal, at best, results obtained in clinical trials attest to this fact. These proteins are very expensive to produce, resulting in the very high cost of therapy. They are not selectively toxic agents. Solid scientific principles explain why IFN's can never be selectively toxic. These proteins produce a huge array of dangerous side effects. Immunologists and molecular biologists argue that these side effects can be diminished, given more money for more research on these agents. Genetic engineering may bring down the cost of producing IFN's, but the use of these proteins can never be justified as anti-cancer agents regardless of cost. The principles of pharmacology predict that IFN's can not enter into most cell types. The principles of biochemistry explain the limited inhibition of the growth rate in cancer types derived from the immune system. To date, you have been forced to spend hundreds of millions of dollars on the development of IFN therapy without receiving any clinical benefit. Using IFN's to cure cancer can never be achieved.

Monoclonal Antibodies

Monoclonal antibodies (MA) are a third protein substance researchers have isolated from the immune system, and another example of the simplistic logic that because cancer is a natural process, another natural process, the immune system, can naturally kill it. This thinking might hold if in fact cancer were a natural phenomenon. Cancer, however, is an unnatural process. Again we see a protein that is suppose to fight cancer is isolated from the cells of the type of cancer it is suppose to fight without any real toxicity to that cell type. The results are predictable. None of the MA's are truly specific for any cell type, as evidenced by the experimental data.

As discussed previously, our immune system attacks micoorganisms. This means that it can recognize self cells from invading cells, such as bacteria and viruses. Immunologists have termed invading cells antigens. In response to this invasion, our immune system produces proteins that can systematically and selectively recognize each type of invader. These proteins are called antibodies. The prefix "anti" means against; so, antibodies attack microbodies, which can be cells, proteins, and DNA fragments.

Monoclonal needs a more detailed definition.The prefix "mono" means one or singular. The suffix "clonal" means exact copies of. Literally translated, the word means exact singular copies of. The term monoclonal antibodies translates into exact copies of proteins which singularly attack a specific antigen. In regard to curing cancer, this means that monoclonal antibodies seek out cancer cells. This translation describes the mechanism of action by monoclonal antibodies.

Researchers are able to take cells of the immune system and, for lack of a better word, "train" those cells to recognize and attack specific cell types. This training is a long, complex and costly process. Briefly, this process involves extracting white blood cells, which are then exposed to the antigen of interest. Those white blood cells that attach themselves to the antigen are harvested because they contain the antibody of interest. The antibody is first purified and, through a complex set of procedures, systematically and rapidly reproduced such that each is the same. This is the process whereby the researcher obtains monoclonal antibodies.

Utilizing this technique, immunologists and molecular biologists have produced a variety of monclonal antibodies, which they claim recognize cancer cells. According to their claims, the MA's become attached to cancer cells selectively and not to the normal cell counterpart. Let us examine the published data to determine if their claims have been substantiated. In one study, researchers developed four different MA's labelled as KM-32, KM-34, KM-52 and KM-93. These MA's were suppose to be very selective against squamous cell carcinoma, adenocarcinoma, large cell carcinoma and small cell carcinoma of the lung. All four of the MA's reacted with some of the cell types, but none of the MA's reacted exclusively with any *one* cell type. Moreover, each MA reacted with the normal cell counterparts of the lung cancer cell types. These reactions hardly meet the criterion for selectivity. In another study, researchers derived a MA labelled EA1 from the human epidermoid lung carcinoma cell line, T222, and fused it with the NS-1 mouse myeloma cell line to attack small cell lung cancer. This MA reacted with 4 different lung epidermoid and adenocarcinoma cell lines. However, the MA also reacted against the following normal cell types: bronchial and alveolar lung, breast, prostatic epithelium, hepatic bile duct, pancreatic exocrine, renal tubules, bladder epithelium and skin epidermis. For a MA that is suppose to react with only lung cells, this MA sure got around in the body. These reactions refute the claim that this MA is selective toward any cell type in the lung.

In another study, researchers isolated MA's labelled NE-25 and PE-35, which were suppose to attack neural tumors and lung tumors, respectively. The MA NE-25 showed activity against a glioma (a type of brain tumor), neuroblastoma (a type of cancer of the adrenal glands), melanoma (skin cancer), renal cell lines, small cell lung cancer, carcinoid (benign tumor) of the lung, thymus tumor, pheochromocytoma (another cancer of the adrenal glands) and the adrenal medulla. The normal cell types that reacted with NE-25 included human fetal and adult brain, nerves, thyroid, adrenal glands and islet cells of the pancreas. PE-35 displayed activity against most lung and epithelium tumors. Normal cells against which PE-35 had activity included bronchial epithelium, epithelium of the gastrointestinal tract, bile duct, pancreas, thyroid, uterine cervix, kidney and skin. Here are two more MA's that attack just about every organ in the body while the investigating researchers maintain that they are selective for a cancer in a given organ. These MA's obviously also fail the criterion for selectivity.

In study after study, researchers report they have developed a MA specific toward a type of cancer when the real evidence consistently runs contrary to their claim.

Knowing the failure of this treatment, yet fighting to save their projects, some of these researchers believe they have stumbled across a way to fool us. There is a technique whereby a chemical can be bound to the MA such that when it attaches itself to tissue, the location can be viewed in the microscope as a yellow-green fluorescent dot. The more yellow-green dots there are in a given location, the more MA's have been attached to the tissue. In the past, researchers using this technique photographed the whole animal. The yellow-greenish dots were seen throughout the carcass.These pictures clearly showed that no MA was selective toward a specific organ. More recently, researchers have omitted pictures of the whole animal in their published articles. Instead, only pictures of the organ they claim was targeted by a MA are shown. The caption beneath the photograph states that this particular MA targeted the organ pictured. These researchers arrogantly believe that withholding evidence can mask their failures.

In reality, MA's target specific proteins. Some of these proteins are pharmacological receptors. Any specific pharmacological receptor can be found in several organs, even those containing cancer cells. It is through this network of receptors that chemical messengers allow various organs to communicate with each other. With this understanding of the mechanism of MA's, it is not surprising to see reactions at several organs throughout the body.

Cancer treatment using MA's has two major insurmountable drawbacks. First, as demonstrated by a post-doctoral research fellow working for SmithKline and Beecham, the production of any MA can not keep pace with the growth rate of any type of cancer. Consequently, the number of cancer cells will always be greater than the number of MA's. Even if there was such an entity as a MA selective for a given cancer, there is no possibility that the cancer can be cured, because there never would be a sufficient supply of MA. The second drawback has a more complex mechanism for failure.

Antibodies recognize and then attach themselves to the cell wall of antigens. Antibodies never enter the cell. After they become attached, a complex series of events is initiated to engulf the antigen by white blood cells, and then to release proteins, such as IL's and IFN's to destroy the antigen. Monoclonal antibodies never enter the cancer cell that they have allegedly been designed to recognize. Attachment of the MA to the cancer cell wall neither kills the cell nor inhibits its growth. The result is that the cancer can not be cured.

Some immunologists admit this limitation in using MA's as a method to kill cancer cells. However, they still believe in the selectivity characteristic of MA's. Of course, this belief means that all they need is more money to develop better MA's through genetic engineering. The result is that you provide hundreds of millions of dollars each year for them to pursue a fantasy.

Some researchers, apparently still believing in the selectivity of MA's but not necessarily their effectiveness, have combined MA's and cancer drugs. In one study, they took an FDA approved anti-cancer agent named methotrexate and chemically bound it to a MA specific for the enzyme prostatic acid phosphatase. This methotrexate-MA conjugate was administered to animals. The researchers reported that more methotrexate accumulated in the tumor than free drug, but significant levels also were found in normal organs such as the liver, heart, lung, blood, kidney and spleen. The toxic effects of methotrexate were reported to be decreased by 10%.

Some researchers will argue that while the results of this study were not as successful as desired, the technique does offer a promising new lead. All they have to do is learn how to refine it. To address this shortcoming, another group of researchers tried a novel approach, which currently is in vogue. Breast cancer in women, as of this writing, is a heated topic. In an effort to target breast cancer cells, researchers derived a MA labelled LICR-LON-Fib75. To this MA, the researchers attached the A-chain from a plant abrin toxin. I do not know what kind of poison this is. I include this study to show you that I was not joking when I suggested that you try any plant extract, animal product or crushed stones you choose to test as anti-cancer agents at home. This article was published in a well respected scientific journal, and supports my contention that many cancer researchers hold fast to the romantic notion that a cure for cancer exists in some exotic plant or animal species, such as the extraction of Taxol from the yew tree. Like Albert Schweitzer want-to-be's, they just need more money to finance their romps in jungles throughout the world to find this "holy grail".

The researchers experimented with this MA-abrin toxin conjugate on women who had breast cancer that had metastisized to the bone marrow. They called this treatment a "cleansing" of the bone marrow. The word "cleansing" strongly indicates that the bone marrow was eradicated of breast cancer cells. Patients who received this "cleansing" went into remission of the disease for an average of 7.3 months. Patients who did not receive the "cleansing" treatment, but who received mephalen, an anti-cancer drug approved by the FDA, remained in remission for 6.2 months. You can tell the physicians believed this to be a very scientific study because they were precise, even down to the first decimal point for the statistical analysis of the months for remission. But, not one woman was cured of her breast cancer.

Removal of the bone marrow is an extremely painful procedure, even though done under general anesthesia. Anyone who has broken a bone can imagine the degree of pain the women experienced. Bone marrow extraction also is an extremely expensive operation. Many insurance companies will not cover this procedure because they consider it experimental. They are correct. Bone marrow extraction for breast cancer treatment is an experimental procedure without any proven benefit to the patient. The fact is that despite all the pain the women must endure and the extreme expense of the procedure, it can not significantly extend their lives. Only physicians and hospitals benefit from expensive bone marrow treatments.

Clearly, all of the studies presented have repeatedly demonstrated the futility of using MA's as anti-cancer agents. None are selective toward tissues in general or organs in particular. In addition, MA's never enter into the cell. Even if a toxin is chemically bound to the MA, that poison will never enter the cancer cell. Even if the toxin could somehow enter the cell, the toxicity of the drug would be substantially reduced because the main reactive center in the drug molecule responsible for the toxic effect will have first reacted with the MA. These are the reasons why the MA-plant toxin conjugate failed to rid the women of the breast cancer that had metastisized to the bone marrow. In essence, the researchers had created a "dud" to be delivered to the cancer cells.

Monoclonal antibodies offer no possibility of curing cancer. No cancer has shown any significant response to this treatment because they are not toxic. Studies have consistently shown that MA's are not selective anti-cancer agents. These proteins are very expensive and time consuming to produce. Their supply can never match the demand due to the faster growth rate of cancer cells. On this research failure, you have been forced to spend hundreds of millions of dollars to date.

Chemotherapeutic Agents

A variety of chemotherapeutic agents with substantial differences in chemical structure are used today in the hope of improving selective toxicity. In Goodman and Gilman's textbook The Pharmacological Basis of Therapeutics, (seventh edition, MacMillan Publishing Company, 1985), a summary of these agents is presented along with a corresponding mechanism of action and toxicity for each. Although the number of drugs is too numerous to discuss in detail, they can be divided into 5 categories based on their mechanism of action: 1. alkylating agents, 2. antimetabolites, 3. natural products, 4. hormones and antagonists of hormones, and 5. miscellaneous agents. Yet with such a large arsenal of drugs at our disposal, cancer still can not be cured. This is because none of these are selectively toxic toward cancer cells.

Alkylating Agents

Virtually all anti-cancer drugs are designed to attack the DNA molecule and to a lesser extent the RNA molecule. The sophomoric reasoning is that DNA synthesis and gene expression will be halted as a result of this onslaught. The tumor will stop growing because cells can not replicate without the DNA molecule remaining intact. Disruption of the DNA molecule also means that genes can not be expressed. The cancer cell is suppose to die because other biomolecules required to sustain life can not be synthesized. The concept may sound logical, but it is incomplete. Chemotherapeutic agents just don't work according to the principles decreed by the researchers.

Alkylating agents are highly reactive chemicals with the ability to bond permanently to the DNA molecule. The presence of large pieces of extraneous material is suppose to prevent enzymes from transcribing the genes to manufacture proteins so that the molecule can be duplicated. Alkylating agents are not explosively reactive, like nitroglycerine, but they are much more reactive than aspirin. A large number of alkylating agents are nitrogen mustards, which are chemical variations of mustard gas. War veterans instantly recognize that mustard gas is another term for nerve gas used in chemical warfare. Many cancer patients are receiving a chemical variation of nerve gas in their veins as treatment for cancer. There is one group of alkylating agents currently in clinical trial that are chemical variations of rocket fuel. These chemicals have an explosion hazard associated with handling them. Still others are variations of insecticides, whose mechanism of toxicity is only partially understood. The alkylating agents are among the most potent poisons ever synthesized by humans. These chemicals indiscriminately kill every life form they contact.

To aid your understanding of the mechanism by which alkylating agents exert their toxic effects, here is a condensed course in organic chemistry. The word organic in this context refers to chemicals made up of carbon and hydrogen atoms, often with some oxygen and nitrogen atoms as well. Many alkylating agents contain either chlorine or fluorine to enhance their reactivity. There is one very important principle regarding chemical reactions. Chemicals have no intelligience. Chemicals can not be computer programmed to react as the chemist, or anyone else, desires . Chemicals can not be trained to seek a target and then react. Chemicals have no conscience; they react whenever the energy environment is favorable. Chemists refer to these principles of reaction as environmental thermodynamics. In short, if the thermodynamic environment allows a chemical to react, it will; if the environment does not favor a reaction, none occurs. This principle applies to all chemicals, including anti-cancer drugs. Chemicals react due to the presence of functional groups in the molecule. Examples of functional groups are the amino ($-NH_2$) and the hydroxyl ($-OH$) moieties. Almost all alkylating agents are designed to react with these two functional groups.

It is logical to ask where the functional groups are located inside of the cell. Each of the four nucleic acids strung together to form DNA, and an additional nucleic acid in RNA, contain amino and hydroxyl functional groups. It is then no surprise that alkylating agents react with DNA and RNA. Whenever alkylating agents come into contact with genes, they will bond, but indiscriminately. No researcher can predict the location of this chemical reaction. Therefore, all genes have the potential to be attacked. Alkylating agents can not distinquish between normal versus cancer cell DNA, despite the fervent desires of cancer researchers. It is important to know that these same two functional groups are found in proteins also. Consequently, whenever alkylating agents come into contact with proteins, a reaction takes place that destroys the function of that protein. Enzymes can no longer properly catalyze reactions necessary for cell life. In addition, structural proteins required for intra-cellular integrity are destroyed as well. These two conse-quences lead to the death of the cell. When enough cells in your body die, so do you. In reality, it is the destruction of proteins that kills cells (and you), not simply the alkylation of DNA.

Alkylating agents react via a non-specific mechanism very much like the reaction of formaldehyde used to preserve organs. These poisons kill all life forms. They can never meet the requirements for selective toxicity against cancer cells.

Antimetabolites

Antimetabolites interfere with the synthesis of genes by substituting false analogues of the nucleic acids into the DNA molecule. This substitution leads to a miscoding of genes so that the appropriate proteins are not synthe-sized. We have a similar situation with these poisons as with the alkylating agents. Normal cells need to synthe-size DNA and have genes expressed as new cells replace those that die out from wear. Antimetabolites inhibit the replication of normal cells as well as cancer cells. These agents are not selectively toxic either.

Natural Products

A number of agents used as anti-cancer drugs have

been isolated from plants and microorganisms. One group of natural products, called Vinca alkaloids, is extracted from the periwinkle plant. Another group of natural poisons, the epipodophyllotoxins, is extracted from the mandrake plant. A third group are antibiotics isolated from fungi of the genus Streptomyces found in ordinary soil. The use of these poisons is another product of the romantic notion among cancer researchers that the cure for cancer is hiding in some exotic life form.

Let's cut through the smokescreen. All life forms have mechanisms for defense. Some life forms can run away. Other life forms are exceptionally strong and use this strength to fight attackers. Most life forms possess neither of these capabilities. They have to rely on manufacturing toxic chemicals for their defense. Because it is extremely difficult to synthesize compounds for each and every type of opponent, plants and some animals synthesize toxins that are very poisonous to many types of potential antagonists, including micoorganism. To maximize their defenses, many microorganisms make poisons that destroy the structure of DNA and other biomolecules. With whimsical reasoning, natural products are sought, extracted and tested against cancers in the fanciful hope that one will be found that extends its toxicity to cancer cells while remaining innocuous to normal cells.

Vinca alkaloids and Taxol disrupt the formation of microtubules required for cell division. The reasoning is that if cells can not divide, then tumor growth is halted. The other natural products disrupt the DNA molecule by either causing strands to break or by inserting themselves into the strands of DNA, which leads to the miscoding of genes and then to cell death. These disruptions, however, are oberved in both normal and cancer cells. None of these agents are selectively toxic to cancer cells, although they are very poisonous.

216

Doxorubicin is a natural product isolated from a species of Streptomyces and is a good example of this toxic mechanism. Doxorubicin intercalates, gets placed between two nucleic acids, into the DNA molecule as a false analogue of nucleic acids. The total dose of this toxin that can be given to a cancer patient is about one gram, or about 1/4 teaspoonful. This is the total *lifetime* dose that can be given to the patient. In addition, nurses must wear gloves to prevent absorption through the skin when administering this toxin so that they do not run the same risk as the pateint of developing irreparable cardiotoxicity. The patient can die from a sudden, massive heart attack as a direct result of being "cured" by using doxorubicin. Other natural products have the same life threatening limitations.

Hormones and Hormone Antagonists

Hormones and antagonists of hormones stop the division of cells by blocking entrance into the cell of the chemical messenger responsible for cell division. This treatment inhibits the growth rate of the tumor, but the cancer cells are not killed. For example, progesterone is given to women with endometrial cancer to the block the entrance into the cancer cell of estrogens that stimulate cell growth. Using a similar logic, Tamoxifen (an antagonist of estrogens) is given to women with breast cancer to block the entrance of estradiol into the tumor cells dependent upon this hormone to stimulate growth.

Using hormones and their antagonists to cure cancer has two major drawbacks. First, none of these drugs kill tumor cells. The growth rate of the tumor merely is slowed. Second, when the hormones are blocked from reaching their primary targets, they are forced to travel to other organs. This blockade from the target organs causes adverse side effects, such as hot flashes, nausea, vomiting, menstrual irregularities, vaginal bleeding and discharge, pulmonary embolism (blood clot lodges in the lungs), thrombocytopenia, and leukopenia (abnormal low white blood cell count).

Miscellaneous Agents

The "miscellaneous" drugs are only chemical variations of alkylating agents, which can never be selectively toxic to cancer cells. All of these poisons are used as anti-cancer agents based upon the same strategy as the alkylating agents of targeting DNA. As discussed previously, this strategy can not possibly be selectively toxic to cancer cells. In reality, cancer researchers have discovered agents that attack all types of rapidly growing cells, but slower growing cells also are killed, just at a reduced rate. Non-cancerous cells die at a slower rate simply because the enzymes responsible for repairing the damage to the DNA have a longer time to function. But these enzymes also can be destroyed by the drugs, leaving the DNA defenseless.

Why Side Effects Occur

All of the cell types in our bodies, except neurons, get replaced, and at different rates depending upon a number of factors such as amount of use and exposure to toxins. White and red blood cells are replaced most frequently, at a rate of about every 3-7 days. Liver cells are replaced relatively frequently because this the main organ where many important biochemical reactions take place, including drug metabolism. For the same reason, kidney cells are replaced at a more frequent rate than cells in most other organs.

Normal blood, liver and kidney cells have growth rates comparable to that of cancer cells. It therefore seems logical to assume that these cells will take up chemotherapeutic agents and be killed at about the same rate as cancer cells. We can predict that the greatest amount of damage should be observed in normal cells of the blood, liver and kidney of cancer patients who receive chemotherapeutic treatments. This is merely an application of the logic used by chemists to target cells based upon their more rapid growth rate than normal cell counterparts.

As predicted, leukopenia and thrombocytopenia are the major limitations in continuing therapy with alkylating agents. These drugs are given intraveneously. If the needle is removed from the patient's vein such that the drug contacts the underlying tissue or the top of the skin, tissue damage will result at that site. This adverse side effect also is consistent with the predicted mechanism of toxicity.

Antimetabolites cause leukopenia, thrombocytopenia, and anemia. Ulcers and localized patches of dead tissue along the gastrointestinal tract, myelopathy (spinal chord damage) and atrophy (wasting away of tissue due to lack of nutrients) of the skin also have been reported in cancer patients prescribed antimetabolite therapy. Other adverse side effects include anorexia, hair loss, nail changes and dermatitis (inflamation of the skin evidenced by itching, redness and skin lesions). These predictable adverse side effects are consistent with large scale, non-selective toxicity.

Bone marrow depression, most often seen as leukopenia, is reported with the administration of vinca alkaloids. Neuromuscular abnormalities is the limiting factor with vinca alkaloid treatments. Other adverse side effects include loss of reflexes, foot-drop, ataxia (muscular incoordination of the voluntary muscles), muscular cramps and neuritic pain. These side effects indicate a general destruction of the peripheral nervous system(the network of nerve cells throughout the body other than the brain and spinal chord). These adverse neurological side effects are consistent with the fact that nerve cells are not capable of replication. Hair loss also is a typical side effect.

Epipodophyllotoxins cause leukopenia, which is the dose limiting toxicity of these natural products. Other side effects include thrombocytopenia, fever, phlebitis (inflamation of the vein, which can be painful), dermatitis, allergic reactions which can turn into anaphylaxis (a hypersensitivity reaction which can cause death), and hepatic (liver) toxicity.

Adverse side effects from antibiotics include leukopenia, thrombocytopenia, anemia and cardio-myopathy (damage to the heart muscle). Severe local toxicities in irradiated tissues of the skin, heart, lung, esophagus and gastrointestinal mucosa also have been reported. Damage to glomeruli (small, sperical structures in the kidney responsible for filtering blood) resulting in renal failure has been reported with antibiotic therapy. As you can see, using natural products to cure cancer offers no advantage over synthetic poisons. Natural products are not selectively toxic agents.

Patients prescribed hormonal therapy for cancer suffer from less severe side effects than with other agents simply because biochemical reactions are temporarily inhibited, but this inhibition can be overridden leading to adverse side effects. Hormones and their antagonists are not especially toxic to cells at the doses administered for cancer therapy because they do not destroy any biomolecules. This relative lack of toxicity observed with hormonal therapy is consistent with the mechanism of action by these drugs.

The miscellaneous agents also cause leukopenia, thrombocytopenia and some neurological disturbances.

The pattern of adverse side effects is similar to that of standard alkylating agents, antimetabolites and natural products because the mechanism of toxicity is the same. With all of the chemotherapeutic agents, a very similar spectrum of adverse side effects is observed, most frequently those associated with the white and red blood cells.

One of the major reasons for the persistent pattern of damage to the blood cells is due to the route of administration of these drugs. Almost all of them are given to the patient intravenously. As a result, blood cells and the bone marrow are the first ones exposed to the poisons. The blood cells consequently take up the largest quantities of the drugs. This leads to the destruction of those cells and the bone marrow. This is also the main reason that there is only limited success against leukemias.

How New Poisons Are Approved for Clinical Trials

In the chapter on home grown cancer research, I described the tissue culture technique, purchasing several cancer cell lines, which did not have to be human cancers, testing arbitraily selected white powders on those cells, and then reporting your results to the FDA in an investigative new drug (IND) application for approval of your white powder in clinical trials. I was not joking about this procedure. Cancer researchers have followed the identical procedure in getting their poisons into clinical trials. They publish their results in journals like The Journal of Medicinal Chemistry, available at most medical school libraries. The following experiments are taken from these scientific publications. All of the compounds described received FDA approval for trial in humans. Keep in mind that all of the information presented so far was available to the researchers who came up with these toxins, and to the FDA officials who gave their approval.

Researchers at Dow Chemical Co. put together a chemical they named Caracemide (CAR. NSC-253272). The scientific name for this chemical is N'-acetyl-(methylcarbamoyloxy)-N'-<u>methylurea</u>. It is another alkylaing agent. I guess the researchers felt that their alkylating agent was better trained to target cancer cells than any similar poisons previously tested. This chemical represents another example of wishful thinking rather than application of basic scientific principles.

There is a well documented chemical carcinogen named N'-nitroso-N-<u>methyurea</u> that has a structure very similar to Caracemide. Some researchers consider this to be the most potent carcinogen known in that it alkylates DNA and causes cancer at extremely low doses in all tissues with which it comes into contact. It is also a very toxic compound. Caracemide alkylates DNA. Why is one chemical a carcinogen and the other an anti-cancer drug? The answer is simply that the researchers who came up with Caracemide wished it to be an anti-cancer drug. Both of these compounds are carcinogens. Even if you know very little chemistry, the chemical name for Caracemide should give a clue as to the carcinogenic potential of this compound. Both chemicals have N-methyurea in their names. It is this methyurea group that is responsible for the alkylation of DNA and the initiation of cancer in the normal cell.

How did the researchers at Dow Chemical Co. get this carcinogen into clinical trials? They tested Caracemide on the F388 lymphatic leukemia cell line and found that it inhibited the synthesis of all large biomolecules, including DNA. The researchers demonstrated to the FDA that Caracemide killed a type of cancer, just as I advised you to do. Remember, I also stated that every chemical at a high enough concentration will kill any life form grown in tissue culture. Normal cells as well as cancer cells synthesize large biomolecules in order to survive. The researchers at Dow never demonstrated any selective toxicity by this compound. This experiment is an example of the misuse of tissue culture as a test for a potential anti-cancer drug. They should have conducted a parallel tissue culture experiment with the normal cell counterpart to determine if Caracemide was toxic only to the cancer cells.

In clinical trials, Caracemide produced neurotoxicity in the form of confusion, disorientation and agitiation. The compound thought responsible for these adverse side effects was methyl isocyanate (a variation of *cyanide*), which formed spontaneously in the body. If this chemical formed spontaneously in the patient's body, then it also formed spontaneously in the medium of the tissue culture. Proper analytical chemistry testing would have detected this poison in the medium. This means that Caracemide is another highly reactive chemical similar to the other alkylating agents. This compound will not explode in your hands, but it will react very quickly and indiscriminately with all tissues. Caracemide is a very potent poison that can spontaneously decompose into another very potent poison inside of the body. Neither toxin is selective toward cancer cells, they are just general protoplasmic poisons that illustrate the myopic vision of cancer researchers.

Two Toxins Seem to Be Better Then One

In another study, researchers came up with a group of compounds named bis (1-aziridinyl) cyclophosphagenes. The prefix "azir" means that these chemicals contain the functional group present in one component of rocket fuel, hydrazine. The suffix phosphagene indicates that these compounds are chemical derivatives of the nerve gas phosgene. I guess that the rationale for combining two dangerous poisons into one molecule is that somehow two poisons are better than one. I have no doubt that based upon the chemical structures of these compounds that they are very potent poisons. I also have no doubt that none of these chemicals are selective toward cancer cells.

These poisons were tested on a mouse leukemia cell line, L1210, and on a mouse lymphoma cell line, L5178Y. Clinical trials with these compounds had to be cancelled due to the cumulative tendency to produce bone marrow toxicity. Cumulative tendency refers to the fact that the chemicals are gradually stockpiled in your body such that in the patient, small portions of the bone marrow are constantly and irreparably destroyed. Continued treatment with these poisons would permanently destroy the patient's ability to produce both white and red blood cells. The concept here is, once again, a footrace between the poison killing all of the cancer cells before the patient dies from the treatment. This race is never won using these chemotherapeutic agents. Notice that the FDA gave its approval to inject this poison into humans without it ever being tested in any human cancer cell lines.

Other researchers have synthesized compounds patterned after the chemical structures of hormones in their effort to target cancer cells. The results with these compounds have been equivocal. Some of the compounds bound more tightly to the hormone receptor, but without any toxic effect to the cancer cells. Other compounds did not bind significantly to the cancer cells, but were toxic to all cell types. There are several flaws in the logic of these researchers. They erroneously assumed that all of the cell types they tried had the hormonal receptor, which was not always the case. Second, if the hormonal receptor is present on the cancer cell, then it is present on the normal cell counterpart. The compound may target the hormonal receptor, but it can not distinquish between cancer versus normal cells. Although the researchers want the compounds to react only with either DNA or RNA, their compounds indiscriminately react with all biomolecules to cause cell death.

We have examined the mechanisms of toxicity for a variety of chemotherapeutic agents. Virtually every agent is very toxic to all life forms. A necessary consequence of the toxic character of these agents is a large array of adverse side effects, with the normal cell types which rapidly divide the hardest hit. The highly reactive nature of these chemicals, along with the front line exposure by the blood cells, explains many of the adverse side effects. None of these poisons are selective for any cell type as the experimental and clinical data indicate. No one of these agents can stop cancer, but this fact does not stop cancer researchers. Ignoring the reasons why these compounds fail individually, they have conducted studies to examine the utility of these agents in combination treatment strategies. This can be compared to using a machine gun instead of a pistol. The machine gun is a more effective killing machine, but it is also less discriminating in what it hits.

CHAPTER 18

WHY ANTI-CANCER THERAPIES FAIL

Because no single chemotherapeutic agent offered any significant benefit to cancer patients, physicians examined multi-agent regimens, which varied in the combination of chemotherapeutic agents and the duration of administration.

Multi-Agent Therapy

The logic for using different drugs was based on the four phases in the life cycle of the cell as it goes from a new to a dividing cell. It was postulated that some agents were more potent at one phase in the life cycle of the cell and that other agents were more potent during other phases. By dosing the patient with multiple chemotherapeutic agents, cancer cells would be attacked at different stages of their life cycle. While some cancer cells would survive one round of dosing during one phase of the cycle, other cancer cells at other phases would not be able to survive subsequent rounds of dosing by second or third drugs. Multi-agent regimens were suppose to cure cancer; so the researchers decreed.

This multi-barreled barrage on cancer cells also was based on the fact that they divide more rapidly than normal cells. Once again we see another strategy to kill cancer cells based upon a simplistic, narrow-minded and unproven understanding of the biochemistry and biology of the cell, while, of course, ignoring the pharmacology. As you now know, almost all of the chemotherapeutic agents are general protoplasmic poisons. We can predict, therefore, that varying the number of toxins and the dosing regimen will not offer any greater benefit than a single chemotherapeutic agent. A multi-regimen therapy strategy also is doomed to necessarily fail. A look at typical published studies will confirm this. These studies are only a very few among thousands reported, but they are representative of the results physicians have obtained.

In one study, 52 patients were treated for breast, liver, sarcoma and ovarian cancers. Sarcomas are a nondescript form of cancer that can arise in almost any organ. Either doxorubicin (a natural product antibiotic) and cyclophosphamide (a nitrogen mustard nerve gas type of alkylating agent) or doxorubicin and vinblastine (a natural product vinca alkaloid antimetabolite) were administered I.V. by a 24-hour continuous pump infusion. A significant number of adverse side effects were observed, including leukopenia, stomatitis (inflamation of the mouth), subclaven vein thrombosis (blood clot formed at the point where the I.V. needle was placed), and 2 deaths resulting from secondary septicemia (a potentially lethal bacterial infection of the blood and tissues). Only 11 patients showed a short-lived favorable response, according to the statistical analysis done by the physicians. None of the patients were cured of their cancers.

A cisplatin (a miscellaneous agent containing platinum), etoposide (a natural product epipodophyllotoxin) and mitomycin (a natural product antimetabolite) regimen was given to 39 patients diagnosed as having non-small cell lung cancer. Of these 39 patients, 22 had squamous cell cancer, 12 had adenocarcinoma and 5 had large cell carcinoma. Adenocarcinoma means that the tumor is a very malignant form of the cancer. Although the side effects of blood toxicities, nausea and vomiting were relatively mild, the average survival time for patients who did not respond well to therapy was 340 days. For the group of patients the physicians labelled as responders to the therapy, the average survival time was 504 days. One group of patients lived about a year longer without a favorable response to therapy, while the other group lived about one and a half years longer due to the therapy. Yet, before the patients were experimented on, the physicians could not predict who would be a responder. This marginal increase in the life span of the patient hardly constitutes a cure of cancer. Of course, the physicians who conducted this study concluded that this treatment regimen offered promising new leads in treating cancer because statistical analysis "proved" the results were significant.

A much larger, international study was conducted by Canadian and French physicians on 295 patients with acute myelogenous leukemia (cancer originating in the bone marrow). The patients were divided into 4 groups according to the regimen they received. Adriamycin (trade name for doxorubicin), vincristine (another natural product vinca alkaloid) and cytosine arabinoside (a nucleic acid derivative of an antimetabolite) were administered in various combinations, with an initial loading dose of the drugs followed by various amounts of maintenance doses.

At the end of this study, they concluded that there was no difference in patient response to the four regimens of maintenance therapy, even though some of the patients received additional hormonal therapy or immunotherapy (IL's, IFN's and MA's) along with the chemotherapy. The authors concluded that it was doubtful that any of the maintenance treatment strategies prolonged survival or remission duration. They based their conclusions on the statistical analysis of the average life span in the four groups of patients; a result we could have predicted from these agents without experimenting on people.

Even though this is just a sampling of all the work published on novel anti-cancer drugs and treatment strategies, you can see a pattern emerging. The strategies employed are based upon the simplistic notion that cancer cells are the only rapidly dividing cells in the patients' bodies, and that damage to the DNA molecule must result in cell death. Neither of these assumptions is true. White and red blood cells divide rapidly and are killed by almost all chemotherapeutic agents. This accounts for the blood disorder side effects reported. Some researchers reported this destruction of the blood cells as the reason for terminating treatment. As you will learn in a later chapter, damage, or alkylation, of the DNA molecule can lead either to the formation of cancer or to no tissue damage.

Gene Therapy

Gene therapy is based on the oncogene hoax. Molecular biologists claim thay can precisely locate and turn off the oncogene, and stop the tumor from growing by reverting the cancer back to a normal cell. However, this genetic alteration is for the future, once they get more money to perfect their techniques. The news media eats up this garbage as though it was manna from heaven. There are a number of reasons why this therapy fails. I already discussed the scientific rationale for the non-existence of oncogenes, as well as how expensive genetic research can be.

A normal gene can not be purchased. It must be synthesized. In order to synthesize the gene, the exact sequence of nucleic acids must be determined. There are from about 1000 to 30,000 individual nucleic acids strung together to comprise a single gene. To locate a gene can take at least five years. Sequencing that gene can take another 5 to 10 years. During this time, tumors continue to grow.

Let us assume that the technology exists that enables researchers to quickly find and sequence genes. Even if the sequence of a particular gene is known, it can not be synthesized from scratch. The technology does not exist to separate a fragment with 100 nucleic acids from one with 101 nucleic acids. As one nucleic acid is added chemically to the larger fragment it becomes increasingly difficult to separate one large fragment from the other fragments. I know how difficult this process is because I worked on a project with the goal of determining the limits of separation for DNA fragments using a more sophisticated technique than the one molecular biologists use. While it was easy to separate 10 nucleic acids from 20 nucleic acids, it was impossible to separate 20 from 21. As the number of nucleic acids increased, the task became increasingly difficult. When the reaction is completed, there can be any number of DNA fragments. It is possible that none of these fragments is the gene of interest.

Molecular biologists claim that they can isolate the oncogene and genetically engineer that gene into the DNA of the bacteria E. coli. They can insert a DNA fragment derived from humans, but there is still the problem of isolating only that gene and harvesting it from the bacterium. They have been able to isolate proteins, such as insulin, from bacteria as a result of genetic engineering, but proteins are easier to separate than DNA fragments, due to substantial differences in their chemical composition.

Let us further suppose that the technology is worked out such that this alleged oncogene can be routinely harvested from either a chemical synthesis process or genetic engineering methods in a reasonably short period of time, and inexpensively. The gene can never be delivered to the cancer cell because the cell wall is designed to keep out large molecules. Genes, which can be larger than monoclonal antibodies, have the same limitations in entering a cell. Molecular biologists would face the same delivery problems with genes as the immunologists faced with monoclonal antibodies. Even if the gene somehow got inside of the cell, it would never reach the nucleus due to the maze of biochemical structures that contain enzymes designed by nature to destroy such large molecules.

Techniques for inserting DNA directly into the nucleus of large cells like ova of mammals have been developed. A sharp, microscopically thin needle is required. Using a special microscope and this needle, a researcher can place the DNA directly into the nucleus. Sometimes the DNA fragment becomes a part of the genetic code. This is a time consuming procedure that works only with large cells. If you have ever done any needlepoint, or sown on a button, you can imagine how time consuming this injection regimen would be. Most cancer cells are so small that this technique can not be used to insert "corrected" genes into cancer patients.

Let us suppose, nevertheless, that this technique was perfected such that insertion of DNA into cancer cells routinely could be done, and that the gene would become incorporated into the genetic code. Suppose that a physician could perform this procedure with needlepoint like precision, and that he could inject a million cells per day. We can safely assume that a solid tumor contains one billion cells. It would take a physician 1000 days to treat the tumor with this technique. The patient would have to be treated every day for three years to destroy just this tumor. The time for locating and treating metastic tumors has to be extended. During this time, it is assumed that none of the tumors will grow, which will never happen. How can they cure leukemias with this technique? Are they going to stick the leukemia cells as they zip by in the blood stream? That would be some mighty fancy shooting if they could. Clearly, the logistics of this procedure are impossible.

Suppose all of these defenses inside the cell are somehow bypassed. Molecular biologists will maintian that with the array of restriction enzymes available to them, they can snip out the defective oncogene fragment and replace it with the correct gene. This technique might be done in the reaction flask, but it is impossible to perform inside of the cell. The restriction enzymes will recognize very specific nucleic acid sequences to cut the DNA molecule, but those exact same sequences are found throughout the whole genetic code. The enzymes can not be restricted to reacting with only that particular sequence molecular biologists order them to. They will continue to seek out all such sequences until none are left. Consequently, all that will be obtained is numerous fragments of DNA with the resulting complete destruction of the genetic code in both normal and cancer cells. Enzymes are incapable of distinquishing normal cell from cancer cell DNA. This strategy presupposes that their restriction enzymes can somehow enter the cell when other similiarly large proteins like monoclonal antibodies can not. It just can not happen.

Some molecular biologists will agree with my logic. However, they will claim that there are "active site directed agents" that can target specific genes. These agents are sent by molecular biologists on a mission to find the oncogene and change it back to the normal gene. This assumes, of course, that the chemical has an intelligience and agrees to accept the mission. You have already learned that any chemical which reacts with the DNA molecule does so indiscriminately; that chemical will react with proteins as well. There is no chemical agent that can target only a gene, and certainly not a gene that does not exist.

Cancer patients were dosed recently with a DNA fragment that constituted a "protective" gene the researchers claimed had inhibited the growth rate of cancer cells in culture. (You have seen how easy it is to produce these results in tissue culture.) With FDA approval, the researchers injected this DNA fragment into the veins of cancer patients. Preliminary results indicated that the patients tolerated this treatment well, i.e., no side effects. Of course the patients tolerated this treatment well because DNA is not toxic. Even if the gene did somehow enter into the nucleus of the cell, there is no guarantee that it would be expressed. Because the concept of the oncogene being expressed and leading to cancer is untenable to begin with, the likelihood that this injected gene will be expressed is non-existent. The gene can not enter into the cell to cause damage. When the white blood cells come across a large molecule like the gene, they will attack and destroy it. The patient suffers no side effects, but the cancer continues to grow. No one to date has reported a cure with this treatment.

I attended a lecture once by a world renowned cancer researcher who presented work on a similar topic. His group claimed that they isolated a gene in E. coli that could produce a protein that would protect humans from the metabolic activation of chemical carcinogens. They placed this gene along with attendant genes to facilitate the expression of the first gene into the nucleus of the ova from a mouse, and then fertilized the eggs. Each of several generations of mice were tested to verify the presence of this alleged protective gene. When these genetically engineered mice were dosed with carcinogens, guess what happened? All of the mice grew tumors. None of them was protected because the bacterial gene was never expressed in the mice. This experiment substantiates my point that just because a DNA fragment in placed inside of a cell, it will not necessarily become expressed. In essence, it is only excess baggage. These researchers still claimed they had made a major advance in understanding cancer prevention. The conceit of these babbling incompetents astounds even me.

A tremendous amount of money is being spent pursuing treatments based on the oncogene myth. This oncogene theory is nothing more than double talk perpetuated by molecular biologists in an effort to have you continue financially supporting their projects involving gene manipulations having nothing to do with cancer.

Gene therapy fails at several different levels. The scientific principles explaining the failure at any *one* level is sufficient justification to stop supporting gene therapy as a viable treatment for cancer. Even if the technological obstacles could be overcome, there are a number of valid reasons based upon the principles of biochemistry, chemistry and pharmacology for the necessary failure of gene therapy as a cure for cancer.

Star Wars Therapy

Radiation therapy is a well known approach to tumor reduction. Patients who receive this treatment suffer many of the same adverse side effects as those exposed to other regimens. The reason for this suffering is that radiation can not target cancer cells any more effectively than the chemical agents used. Researchers have developed variations of radiation therapy in efforts to enhance the targeting ability of the treatment for cancer cells, and thereby reduce the incidence of adverse side effects.

One of these variations is called the gamma ray knife. Gamma rays are a form of radiation similar to x-rays. The principle guiding the justification for this gadget is that surgeons can more accurately focus the radiation beam on the tumor. In this way, only tumor cells are suppose to be excised, so the patient suffers fewer side effects. Another closely related variation of this toy is the computer controlled laser beam. Surgeons can lock onto a set of coordinates for the tumor, in Star Trek style, and zap out the cancer cells. These are interesting ideas, but what happens when the patient moves, even breathes? The coordinates the surgeon locked onto will change; normal cells will be killed instead of cancer cells. Death of normal cells leads to adverse side effects. Also, the surgeons face the same logistics of needlepoint accuracy as with gene therapy.

Using these electronic novelties offers no advantage over existing treatments. Surgeons using either the gamma ray knife or the laser must still face the problems of possibly cutting major blood vessels and nerves, just as with the scalpel. Even with these toys, there is no guarantee that all of the tumor will be removed. They are useless against leukemias. Cancer can not be cured with these gadgets.

Each one of these gadgets costs about $1 million. Your increased costs for hospital care pay for the bragging rights of hospitals and physicians for the newest toy in the medical products market and the latest technology in cancer treatment. They are not lying because this is the most advanced technology, although nothing more than electronic scalpels. But they will never admit the truth that these new gadgets offer no innovative advancement beyond a scalpel for curing cancer.

TNF Therapy

Some venture capitalists are investing heavily in the genetically engineered protein isolated from the natural killer cells of the immune sytem. Again, another protein isolated from the immune system. Going back to nature in cancer research never seems to go out of style. Researchers named this tumor necrosis factor, or TNF. Necrosis means the death of body tissue, but not the whole body. The researchers ordered TNF to kill only and all cancer cells. To determine this killing capacity, the researchers tested TNF against 6 human breast and bowel cancer cell lines. These tumors were grown in "nude" mice. Nude here means that the mice had severely impaired immune systems.

After a period of time sufficient to allow the tumors to grow, the researchers injected the TNF by one of two methods. In the first, TNF was injected into the blood-stream of the mouse. None of the 6 cancer cell types responded; all continued to grow. In the second method, TNF was injected directly into the tumor. Only 3 of the cell types responded, but the tumors did not disappear. Yet, the researchers claimed they discovered a promising new lead to cure cancer. Let us examine the logistics of this treatment. According to the results reported by the researchers, TNF has to be injected directly into the tumor to have some success in killing cancer cells. The physician would have to find all of the tumors and inject each one with TNF. No single injection of TNF will kill all of the cancer cells. This treatment also becomes reduced to a footrace between surgical technique and tumor growth rate. Some metastatic tumors can easily go undetected, which means that a small tumor has a chance to grow around a major blood vessel or nerve, rendering it inoperable before the patient has any symptoms. Like gene therapy and the laser beam, just the act of injecting TNF into varous sites on the tumor is time consuming, and expensive. All of this activity presupposes that the cancer cell type will respond when even the researchers reported no response by half of the cancer types they tested. This experimental treatment can cost you $25,000-$75,000 with no guarantee that your cancer will be cured, or even the growth rate inhibited. Because this is an experimental procedure, it is unlikely that your medical insurance companies will cover the cost. These logistical problems do not address the greater difficulties faced with leukemias. Leukemias can not be cured using TNF because it would have to be injected into the blood, subjecting it to attack by the immune system; unless, of course, the physicians are sharpshooters.

Even the researchers who reported this experiment did not know the mechanism of action explaining the toxicity of TNF. Because they proclaimed it a successful treatment for cancer, molecular biologists are busy trying to find the gene coding for TNF. This gene will then be genetically engineered into E. coli for large scale manufacture, like insulin. The price tag for these genetic engineering experiments, according to the March 2, 1992 issue of Business Week, can be estimated at around $250 million. If the experiments are ever completed, cancer researchers will have another useless protein by which to victimize unsuspecting cancer patients.

The Sad Conclusion

We have examined a variety of typical anti-cancer agents and treatment strategies. All have failed. Interleukins, interferons, monoclonal antibodies and tumor necrosis factor are proteins isolated from our immune systems. The rationale for using these proteins is based on the decree that our immune system is a natural process that can be "trained" to destroy another natural process called cancer. Chemotherapeutic agents and gene therapy focus on the alteration of the DNA molecule to kill cancer cells. Some anti-cancer drugs are natural products isolated from plants and microscopic life forms. The use of these agents is based upon the simplistic notion that damage to the DNA molecule leads directly to the death of the cell. Chemical carcinogens also damage the DNA molecule without producing cell death. Self-proclaimed medicinal chemists seem incapable of recognizing the contradiction between alkylation of DNA leading to cancer and their wish that their alkylating agents will cure cancer. Using million dollar electronic gadgets like the gamma ray knife and the surgical laser is a scam run by surgeons to get large fees for playing video games inside of their patients. Hospitals capitalize on this because it is relatively cheap advertising that brings in huge profits.

All of these anti-cancer treatments are experiments, and you are the guinea pig. None of these treatments are based upon solid scientific principles. As I described in the chapter on duck hunting, cancer researchers are shooting in every and any direction in an attempt to find a cure for cancer. As they become more desperate to secure funding, their methods become more bizarre and expensive. Cancer researchers lack the requisite expertise in the field to which they lay claim. This is one of the main reasons cancers have not been cured. Those researchers use any method to intimidate and impress you into believing that they are making real progress. Their publications and press releases verify that they are terrible, life threatening failures.

But despite these obstacles, cancers can be cured.

CHAPTER 19

THEORY AND DICHOTOMIES

Webster's dictionary defines the word theory as, "A formulation of apparent relationships or underlying principles of certain observed phenomenon which has been verified to some degree." This definition can be interpreted as another word for opinion. A theory has to bring together principles, substantiated by data, that explain the relationship between sets of data such that future events can be predicted, at least qualitatively. Ideally, future events are then quantitated through experimentation. A theory also has to be able to explain the relationship between any apparent dichotomies that arise. Current theories on carcinogenesis may meet the criterion of Webster's definition, but not mine.

Theory and Opinion

Let me illustrate the correct application of the word theory using the example of magnetism. Suppose there are two individuals with two bar magnets each. If all either of them did was to put two ends of the magnets together, they could each report a different result. One would insist that the magnets only repel each other, and the other would maintain that they only attract each other. If they were like typical cancer researchers, they would spend millions of your dollars and a lifetime arguing over who's right and who's wrong without ever settling the issue. In reality, both are 50% right and 50% wrong.

To settle this dispute, a theory on magnetism would have to be developed to explain this dichotomy. This theory would establish guiding principles to predict the effects of a magnetic field. We in fact know why magnets repel or attract each other. Numerous experiments have been successfully conducted to verify the theory on magnetism. Mathematical equations have been developed that describe the effects of magnetism. Books have been written to enhance our understanding of magnetism. Courses are offered in universities solely on this subject. Many of the high-tech products we enjoy today and will in the future are based on the application of magnetic theory. This theory works and it is used to design products. A theory on the mechanism of carcinogenesis must be as productive.

It has been very difficult for scientists to develop a rational theory on carcinogenesis because the methods of data collection are shoddy, the interpretation of the data is misguided; usually its for both reasons. You learned that this failure is due to the egotistical tyrants who refuse to accept as valid any information or explanation that contradicts their pet theories. Their theories are little more than misguided opinion.

Criteria For Valid Data

What constitutes good data? What is the criteria for valid interpretation versus opinion? These are difficult and important questions, but ultimately, as grandma once said, "The proof is in the pudding". I use a checklist of questions for any given set of data, especially mine. There also is a subjective test I use, which Einstein referred to as gedanken experiments. This is a thought process consisting of a series of conditional statements that test the validity of each logical consequence of the statements. If an explanation of the data is valid, then the logical consequences should be an outcome predicted by the explanation. An experiment is then designed that tests the explanation and the validity of the consequence. Even if the outcome is negative, the experiment still has value in that it can point the scientist in the right direction with a greater understanding of the mystery. Very few scientists are capable of performing gedanken experiments. This is one of the reasons most cancer research is an expensive exercise in futility.

My checklist is as follows.

1. Is the data reproducible?

Reproducible means that several researchers have reported essentially the same results. It also means that any researcher can obtain the same results everytime, if the same procedure is followed as described in the report.

2. Is the data logically consistent?

The data collected should be appropriate with the goal of the experiment. For example, Lake and Cottrell knew that the metabolism of some nitrosamines was only partially accounted for by the P-450 enzymes from rat liver microsomes. They hypothesized that monoamine oxidase (MAO) had to be responsible for some of the metabolism due to the structure of these compounds. To test their hypothesis, they designed experiments to measure nitrosamine metabolism by MAO. When they used a number of MAO inhibitors, the metabolism of the nitrosamines was inhibited, as they had hypothesized. Both sets of data pointed to MAO metabolism of some carcinogens. Their data was consistent with the goal of the experiment. In contrast, the researcher who reported that she measured the expression of oncogenes in all tissues after dosing the animal with a nitrosourea that produced tumors only in the pancreas and liver of the animal had inconsistent data. Results consistent with her logic and theory would have been oncogenes expressed only in the pancreas and liver but *not* in any of the other organs, or that tumors formed in all the organs that expressed the alleged oncogene. Who knows what she really measured.

3. Is the methodology for data collection logical?

Proper use of instruments to collect the data and, if applicable, statistics to analyze the data have to be applied at all times. Again, there has to be an inherent logical consistency. For example, I discussed the biochemist who used the spectrophotometer to forever dismiss the possibility that MAO metabolizes nitrosamines. Following his procedure, it is impossible to stabilize the instrument; he could not have obtained the data he claimed. I know this to be a fact because I tried the experiment as he described it. Likewise, the researcher who allegedly "proved" that high voltage power lines caused a 1.7 increase in the incidence of cancer in children improperly used statistical analysis. Statistics prove nothing. They serve as a guide that indicates how to proceed for further understanding of a subject.

4. Is the data collected consistent with established, <u>valid</u> theory?

I hope that I have not given you the impression that no good work, strong theories, or fine scientists exist. Excellent work has been done by strong minds who have established solid principles in all of the sciences. It is just that this type of scientist generally has been ostracized from cancer research. Again, let me use the work of Lake and Cottrell as an example. They never challenged the contention that the P-450's act upon nitrosamines. But they logically reasoned that because only 35% of the metabolism could be accounted for by this system, some other enzyme system(s) must be active. They applied principles expounded by the theory of structure-activity relationships to reason that MAO must be one of these enzymes. Their data supported their hypothesis. This contrasts with the oncogene hoax. There is no logical genetic theory that predicts the existence of these alleged genes. In fact, genetic theory and measured data strongly argues against this fallacy. However, the decree was made and the subordinates were ordered to manufacture data to support it.

When I read other researchers' work, I always ask, "Is the collection of data and its interpretation logically consistent with established theory?" I use this same criterion on my work. One trap to avoid is accepting dogmatic conditional statements. For example, if X exists then Y must have caused it; or if Y is present then X must result. X and Y are separate events that may or may not have a causal relationship. Often times, the two appear together, but they may also appear without each other. This is the type of situation that arises when physicians misuse statistics in cancer research to link an item to a form of cancer. Molecular biologists argue that cancer is genetically determined because they supposedly found a fantom oncogene, although they still can't say what turned it on.

Dichotomies

In developing the Hegedus Theory on the mechanism of carcinogenesis, a number of dichotomies in the literature had to be addressed. Any theory that can withstand the test of time must explain how and why these contradictions co-exist while incorporating all data that is reproducible and valid. In addition, any new theory must present principles that can guide researchers to predict future events more precisely than current opinions. There are a number of problems in explaining carcinogenesis with any of the currently popular theories.

Alkylation of DNA

Alkylation of DNA is generally accepted as a mechanism of carcinogenesis. However, in experiments with the same type of animal, every combination of high and low concentrations of alklyated DNA has been reported as detected in both target and non-target organs after dosing with chemical carcinogens:

1. High concentrations of alkylated DNA in the target organs, with low concentrations in non-target organs.

2. Low concentrations of alkylated DNA in target organs, high concentrations in non-target organs.

3. High concentrations in both target and non-target organs.

4. Low concentrations in both target and non-target organs.

To verify that the alkyl moiety came from the carcinogen, chemists have used either a radioactive or a non-radioactive isotope of carbon or hydrogen to tag it at a specific location on the alkyl portion of the molecule. A chemical fingerprint of alkylated DNA using the mass spectrophotometer verified that the alkyl moiety came from the tagged carcinogen. The amount of alkylated DNA has even been quantitated with high performance liquid chromatograph (HPLC). I also have conducted this kind of experiment and have obtained the same results. The DNA in these experiments was unquestionably alkylated by the carcinogen, but there has never been an explanation for how every combination of site and level of detection listed above can occur. Moreover, some chemical carcinogens, like the formerly prescribed anti-histamine methapyrilene, do not alkylate DNA. Still other compounds reported as carcinogens, like the formerly prescribed antipsychotic drug reserpine, exert their actions at the cell's membrane and never enter the cell. No current theory on carcinogenesis can even begin to explain, let alone reconcile these conflicting results.

Potency Versus Rate of DNA Repair

The order of potency for a series of nitrosamines is, diethylnitrosamine (DEN) with 2 carbon atoms in the alkyl moiety is slightly greater than dimethylnitrosamine (DMN) with 1 carbon atom in the alkyl moiety which is greater than dipropylnitrosamine (DPN) with 3 carbon atoms in the alkyl moiety which is very much greater than dioctylnitrosamine (DON) with 8 carbon atoms in the alkyl moiety. Metabolism of DEN forms O^6-ethylguanine and O^4-ethylthymine on DNA; DMN forms O^6-methylguanine and O^4-methylthymine; and DPN forms O^6-propylguanine and O^4-propylthymine. DON does not form alkylated DNA adducts. In general, as the side chains (number of carbons) in the nitrosamine increase in size, potency as a carcinogen decreases. In fact, DON is not a carcinogen.

There is an enzyme called de-alkylase that patrols the entire genetic code seeking to remove alkyl moieties attached to oxygen atoms on guanine and thymine. The order of affinity, or rate of repair, for the alkyl moieties is, methyl greater than ethyl which is much greater than propyl. As the size of the side chain increases, affinity for the de-alkylase enzyme decreases. This means that the longer the alkyl chain, the longer it remains attached to the DNA molecule. If alkylation of the DNA molecule is the precipitating event for carcinogenesis, to "switch on" the oncogene as claimed, then the nitrosamines with the larger alkyl moieties should be the most potent because they remain longer on the nucleic acid, enabling it to initiate carcinogenesis. The fact is that potency of a carcinogen runs contrary to the length of time the alkyl moiety remains attached to DNA.

Another mystery reported in the literature is that no alkylated DNA adducts are detected more than 72 hours after the last dose by a chemical carcinogen. This is true in both target and non-target organs, and regardless of which chemical is used. Alkylated DNA adducts are not necessarily present when tumors form.

Onset of Cancer

It seems that some people never expose themselves to carcinogens, yet they develop cancer. Others seem to imbibe carcinogens, but do not develop tumors. Use of tobacco products is a good example. Many patients never smoked, but they developed lung cancer. At the same time, some three pack a day smokers never develop lung tumors, but they often develop another debilitating disease of the lung called emphysema.

It is easy to understand how a heavy smoker can get lung cancer in a few years. However, sometimes a person is exposed to just one relatively high dose of carcinogen and a tumor is found twenty to thirty years later, seemingly all of a sudden. A large number of women are perplexed because they get mammograms routinely every two years for the first signs of breast cancer. For perhaps a decade, no tumors are detected. Then it seems that suddenly a large tumor is present from nowhere. The woman did everything right according to the best medical advice, yet she still got cancer

In an organ, two cells can exist literally side by side with one becoming cancerous while the other remains normal. Oncogene theory certainly can not explain how this dichotomy can exist. In fact, this theory argues that not only these cells, but all of the cells in the organ must become cancerous simultaneously.

Current theories can not explain either the delayed or the rapid onset of cancer. Nor can they explain the varying responses by different cell types in the same organ leading to cancer formation, let alone the variable carcinogenic response by the same cell type.

Radiation

As discussed previously, virtually all chemical carcinogens target organs depending on the specific carcinogen's chemical structure. The most notable exception to this rule is methylnitrosourea which is a contact carcinogen in that it causes cancer in every organ it touches. Radiation is not organ specific, but it can appear to be.

To understand the mechanism of carcinogenesis from radiation, a short course on this subject is necessary. There are three majors forms of radiation: alpha particles, beta particles, and gamma rays. X-rays and ultraviolet light can be considered variations of gamma rays. Alpha's are relatively large subatomic particles that move a very short linear distance, about three inches, but they can cause a great deal of tissue damage within this range. For tissue damage, or cancer, to result from alpha bombardment, the source has to be ingested, like inhaling radon gas, which is an alpha emitter. Beta particles travel in a linear path up to about 6 feet. Like alpha particles, they too usually must be ingested to cause tissue damage and cancer, but they cause less damage. Carbon-14 and tritium (hydrogen-3) routinely used in medical research are beta emitters. Both are natural isotopes of their respective elements. Generally, we are protected from alpha and beta particles by clothing and distance from the source.

Gamma rays can travel in a straight path the length of the universe at the speed of light. They are extremely small. Ingesting this form of radiation usually does not cause damage because the rays are so small they can pass through the body without hitting any biomolecules. Their greatest danger comes from external sources, like the sun and x-ray machines. If a gamma ray hits a biomolecule, it can cause the same kind of damage as the other two forms of radiation. Lead shields are used for protection from gamma rays.

Because we can not predict the angle of the paths any given particle of radiation will take, no one can know which organ(s) will be hit. Further, it can not be predicted when a radioactive atom will decay to emit an alpha or beta particle, or a gamma ray. Consequently, the radioactive atom can come into your body and be eliminated without ever decaying to cause damage. On the other hand, it can circulate in your body and decay in an organ that was not the route of entry to cause tissue damage, including cancer.

For example, radon gas usually causes lung cancer because this in the organ for route of entry. If the radon atom does not decay before being exhaled, no lung tumors will form. If radon decays in the lung, then there is a good chance that lung cancer will develop. If radon enters the bloodstream from the lung, it can go to another organ, such as the liver, and decay to cause damage but no cancer. As a general rule of thumb, the organ closest to the source of radiation is the most likely to develop cancer.

From a high exposure to radiation, a person may develop cancer or not develop it even with severe tissue damage throughout the body. In contrast, a person may be exposed to very low levels of radiation, such as an x-ray for diagnosis of breast cancer, and develop tumors.

Carcinogenesis Versus Mutagenesis

A large number of chemicals that cause cancer also produce mutations, genetic alterations, in several species of animals and microorganisms. However, a chemical can be a carcinogen, but not a mutagen. Conversely, a chemical can be a mutagen without being a carcinogen. This is especially puzzling because carcinogens cause damage to the DNA molecule, and damage to genes ordinarily leads to mutations. No one has been able to offer a satisfactory explanation for this phenomenon.

Numbers and Luck of the Draw

All of the above dichotomies are explained by the Hegedus Theory. To understand this theory, keep in mind two guiding principles. One, carcinogenesis is primarily a question of numbers. Low carcinogen concentrations, or low numbers of carcinogen exposure do not produce cancer. High concentrations, or exposure, produce substantial tissue damage, including cancer. And two, cancer results from a chance encounter between the carcinogen and the cell, or luck of the draw.

CHAPTER 20

THE HEGEDUS THEORY OF CARCINOGENESIS

To understand the mechanism of carcinogenesis, it is essential to learn how normal cells divide. Remember, no gene becomes active until it receives its messenger. Once all of the biomolecules have been manufactured and put into place, there is very little, if any, genetic expression. Normal cells do not divide until a very specific chemical messenger, different for each cell type, instructs them to. A particular series of biochemical events must occur before the message is received by the nucleus.

Deactivating Enzymes

It is my belief that there are no enzymes made specifically to metabolize foreign chemicals. Rather, there are enzymes designed to metabolize endogenous (made inside the body) chemical messengers, like neurotransmitters and hormones, and nutrients, like glucose. An enzyme will metabolize a foreign chemical only if that chemical closely resembles the endogenous chemical for which the enzyme was designed. The enzyme will make the same change in a foreign chemical's structure as in the endogenous chemical. If the metabolite of the foreign chemical has a structure similar to a second endogenous chemical or nutrient, then a second enzyme will metabolize it to form the same corresponding chemical change as in the second endogenous chemical. This sequential metabolism continues until a final metabolite is passed out from the cell.

As long as the intracellular (within the cell) concentration of the foreign chemical remains low, no life threatening changes occur to any of the enzymatic pathways inside the cell. Some reactions may be slowed, but not to an appreciable extent. The cell does not "sense" danger. However, as the amount of foreign chemical increases, the first of two intracellular critical concentrations of endogenous chemical messenger is reached to signal danger. If the enzymes in the cell are occupied with the metabolism of the foreign chemical, they can not metabolize, or deactivate, the endogenous chemical messenger that stimulates gene expression. At the first critical concentration of endogenous messenger, it either binds to an appropriate gene, or to a protein receptor to form a messenger-protein complex that in turn binds to the gene. This stimulates the production of RNA coding for the enzyme which deactivates the endogenous chemical messenger and metabolizes the foreign chemical. The challenge to the well being of the cell is successfully met.

At still higher concentrations of foreign chemical, the cell can not produce enough enzyme to metabolize it. Also, the cell has room for only a limited number of enzymes. At still higher concentrations of foreign chemical, it occupies the deactivating enzyme such that none of the endogenous chemical messenger is metabolized. Eventually, the second critical concentration of messenger is reached which signals that the cell is in mortal danger of being killed. To meet this challenge, the cell divides. With two cells, the amount of enzyme is now doubled and sufficient to deal with the foreign chemical.

Again, the challenge is successfully met. Should one of the two progeny cells die, the other cell can replace the functions of the parent cell.

In addition to stimulating the production of an enzyme for deactivation and cell division, the messenger stimulates the expression of genes that code for proteins which provide the specialized function of a given cell type, a process called differentiation. Most often, endogenous chemical messengers are hormones and neurotransmitters. For example, the hormone estradiol stimulates both breast enlargement and the production of special cells in the breasts when girls reach puberty. This hormone is also instrumental in regulating the menstrual cycle, including the temporary slight breast enlargement during the secretory phase of the cycle.

Cell proliferation and differentiation is a natural and continuous process that ensures the life of every organism. This process is in operation to constantly replenish the organ with new cells as old cells wear out and die. When the cell is confronted with high concentrations of a toxic foreign chemical, this defense mechanism rapidly and efficiently moves into action to destroy the invader.

Biochemists and pharmacologists take advantage of this mechanism when they study the metabolism of a new compound that shows potential as a drug. They routinely pretreat animals with various chemicals that induce the formation of higher than normal amounts of specific enzymes to measure the way they metabolize the new compound. This is an important procedure because many times the parent compound is relatively nontoxic, but one or more of its metabolites can be lethal. For example, phenacetin was sold over the counter at one time as a substitute for aspirin. Eventually, some people died from using this drug. It later was discovered that acetominophen (Tylenol®) is the active metabolite that produces pain relief while another metabolite alkylated proteins in the liver, killing the user.

Cell Division

Before a cell divides, a necessary sequence of biochemical events take place. First, the enzymes that transcribe DNA into RNA are put into action. The RNA then leaves the nucleus to initiate the production of proteins in an organelle called the ribosomes where enzyme synthesis takes place to form the appropriate amino acid sequence. These first few enzymes are probably the ones required for the production of energy and a corresponding pharmacological receptor for that cell type. Next, a DNA polymerase is activated to synthesize a duplicate of the genetic code. Now, the cell is ready to divide.

After the cell divides, a second sequence of events occurs to ensure the survival of the cell. Proteins that enable the cell to perform its specialized function are manufactured. Structural proteins are put into place and enzymes that help the cell to defend itself are manufactured. Finally, in the last phase of the cycle, the specialized structures in the cell, like the mitochondria, the microsomes, and the endoplasmic reticulum are erected. These biomolecules are synthesized in decreasing order of importance to the survival of the cell.

There is a logic to this sequence. Constructing the huge biomolecules, like DNA, RNA and proteins, takes a large amount of energy; therefore, enzymes that convert glucose into energy must be ready. The complete set of DNA must be synthesized in duplicate, or one of the new cells will lose part of the genetic code. The receptors are necessary to uptake the appropriate messenger while excluding all other messengers as a means to ensure that the two progeny cells differentiate into the appropriate cell type. Otherwise, there would be mass biochemical confusion in the cell. Cells can survive without some organelles, the mitochondria for example. In fact, cancer cells derived from tumors in the advanced stages of growth have very few mitochondria. The cell also can survive without some structural proteins that construct a set of organelles called the endoplasmic reticulum, which contain the P-450 enzymes and the enzyme that can deactivate the mesenger.

To sum up the process, the intact cell does not divide until an appropriate endogenous chemical messenger specific for that cell type reaches a critical concentration in two steps. Once this condition is achieved, the cell divides according to a hierarchical ranking based on the importance of the biomolecule to the survival of the cell. The most important biomolecules are made first; the least important are made last.

Only two types of carcinogens exist: chemicals, and radiation. The Hegedus Theory is based on chemical carcinogenesis with slight modifications to include radiation. Chemical carcinogenesis is discussed first, followed by carcinogenesis due to radiation. This theory is simply an application of the theory for normal cell division with an explanation of what went wrong to produce cancer.

Chemical Carcinogenesis

Contrary to popular belief, cancer cells do not crop up like mushrooms on the lawn in Spring. Only certain cell types can become cancerous. Other cell types may be present in the tumor, but they are not cancerous. A cell with the potential to become cancerous must possess four characteristics.

These four characteristics are:

1. The cell must be able to synthesize and secrete an endogenous chemical messenger.

2. A corresponding pharmacological receptor must be present on the cell membrane to selectively re-uptake it.

3. The cell must synthesize the enzyme that is specific for the deactivation of this messsenger.

4. This endogenous chemical messenger stimulates gene expression, including cell division in that cell type.

These four characteristics comprise an elaborate system for the cell to sense and react to life threatening danger.

Chemical carcinogens target a specific cell type in an organ depending on how closely their structure mimics a given endogenous chemical messenger. This similarity in structure "fools" the pharmacological receptor into selectively uptaking it as a false analogue of the messenger. Once inside the cell, the same deactivating enzyme specific for the messenger also metabolizes the carcinogen to produce one or more reactive metabolites.

If the concentration of carcinogen is low, then the activity of the deactivating enzyme may be destroyed but the gene coding for it will remain intact. Any damage to DNA, like alkylation, can be repaired before the cell divides. The intracellular amount of the messenger will increase to the first critical concentration, stimulating the gene coding for the manufacture of more enzyme. This additional enzyme will then metabolize the additional carcinogen, which will be harmlessly removed, returning the cell to normal function.

If the concentration of carcinogen is relatively high, again more enzyme will be manufactured to meet the challenge by the same defense mechanism, but the enzyme will be quickly rendered nonfunctional due to the production of reactive metabolites which permanently bond at several amino acid sites along the chain of the enzyme. Consequently, the second critical concentration of messenger is reached, and the cell divides. Again, as long as the DNA did not become alkylated or it was repaired in time, no alterations in the genetic code will occur. The cell will not become cancerous.

For the cell to become cancerous, three conditions must be present simultaneously.

1. A portion of the DNA molecule must be damaged, usually by alkylation at O^6-guanine and/or O^4-thymine.

2. This damage must be at a key location on the gene coding for the deactivation enzyme such that the sequence of amino acids necessary for the proper function of the enzyme is permanently altered.

3. The second critical concentration of messenger has been reached, stimulating the cell to divide.

Because the parent cell did not have the time to repair this damage, the two, and all subsequent, progeny cells will permanently contain this mutation on the gene coding for the deactivating enzyme, and it will be passed on to all future cells. None of these cells will ever be able to synthesize the enzyme that deactivates the endogenous chemical messenger. In other words, these progeny cells have lost the switch to turn off the signal that initiates cell division. They now are cancer cells.

It is important to realize that the chemical carcinogen no longer has to be present. It did its damage. The chemical carcinogen only has to be present long enough to cause the damage. After that, the series of events takes off on its own momentum.

This series of events in these new cancer cells is as follows. The cancer cell still synthesizes and secretes the same amount of endogenous chemical messenger as its normal cell counterpart. Also, like its normal cell counterpart, it selectively uptakes this messenger at the same rate. Unlike its normal cell counterpart, however, it can no longer deactivate this messenger. The intracellular concentration of messenger continues to increase. When the first critical concentration is reached, the gene coding for the enzyme no longer is present for the messenger to bind. Or, the messenger can bind to the same location on the gene that once coded for the enzyme only to stimulate the production of a nonfunctional protein instead of the deactivating enzyme. Or the messenger could migrate to the next position in the DNA that codes for cell division. The second critical concentration may not have to be reached before the cell divides. Overall, the time lag between cell divisions is decreased. To put it another way, the growth rate of the cell has been accelerated. This cycle of events continues, eventually producing a large mass of cells called a tumor. From only one cell that had genetic damage which could not be repaired in time, a tumor results.

Cellular growth rate is dependent on the concentration of messenger. As more cancer cells form, more messenger is produced accelerating the growth of the tumor. It is important to understand that each individual cell does not synthesize and secrete a larger than normal amount of messenger. The increase in messenger concentration is due to the larger number of cells which can not deactivate it. As time goes on, the cancer accelerates its own growth rate.

Carcinogenesis is a question of numbers. Low concentrations, or numbers, of carcinogen do not produce any adverse effects. Increasing the amount of carcinogen increases the chances of producing permanent genetic damage that is passed on to progeny cells. Increasing the concentration of endogenous chemical messenger stimulates the cell to divide. Nothing happens below this concentration. When the number of cancer cells in the organ is low, the growth rate is not substantially increased because the concentration of messenger is low in this initial phase of cancer. Gradually, the number of cancer cells increases leading to more messenger that accelerates the growth rate, increasing the size of the tumor.

Each cell in the body contains the same DNA for the complete genetic code for every protein synthesized in the cell, including enzymes. The DNA molecule exists as a double helix of two complementary strands. Each nucleic acid is bonded to the next one by a sugar-phosphate linkage. The four nucleic acids, adenine, cytosine, guanine, and thymine, are joined together along the strands by hydrogen bonds that protrude into the axis of the helix to form complementary base pairs. Adenine only bonds to thymine, while cytosine only bonds to guanine. When the cell is ready to divide, the two strands separate so that two new complementary strands can be synthesized to make two new exact replicate and complete sets of genes. Each of the two new cells retains one of the new sets of genes.

A minute change in the nucleic acid sequence of a gene can yield drastic alterations in the structure, and hence function, of the corresponding protein, especially if that protein is an enzyme. The double helix of DNA is formed by the pairing of guanine on one strand to cytosine on the other via three hydrogen bonds, one of them at the O^6- site; and thymine on one strand to adenine on the other via two hydrogen bonds, one of them at the O^4- site. When polymerase moves along the DNA molecule to duplicate a gene, it reads a three point attraction when the nucleic acid is either guanine or cytosine. It also can read if the nucleic acid is aromatic, as is cytosine, or non-aromatic, as is guanine. Thymine, a non-aromatic nucleic acid, forms two hydrogen bonds with adenine, an aromatic acid. The aromatic nucleic acid pairs with a non-aromatic nucleic acid. Based on these two readings of the nucleic acid on the template DNA strand, the enzyme selects the proper nucleic acid to elongate the sequence on the replicate strand. For example, when polymerase reads three hydrogen bonds and non-aromatic nucleic acid, it "detects" guanine; therefore, its complementary nucleic acid, cytosine, will be added to the other strand during gene replication. If polymerase reads two hydrogen bonds and aromatic nucleic acid, it detects adenine and adds thymine to the replicating strand.

When guanine is alkylated at the O^6- position of the template strand, it can not form three hydrogen bonds with the complementary nucleic acid, cytosine, due to the steric (spatial) hinderance by the added alkyl moiety. As polymerase moves along the template strand of DNA to synthesize the replicate strand and it meets O^6-ethylguanine, for example, it reads a two hydrogen bond, nonaromatic nucleic acid, which it interprets as thymine. Instead of adding cytosine, guanine's complement, to the replicate DNA strand, polymerase adds thymine's complementary nucleic acid adenine. In the replicate strand, a three hydrogen bond, aromatic nucleic acid has been replaced by a two hydrogen bond, aromatic nucleic acid.

During the process of cell division, the replicate strand of DNA serves as a guide for the duplication of the originally mated strand of DNA. Polymerase reads the replicate strand to synthesize the new complementary strand. At the location of the mutation, the enzyme reads adenine in the replicate strand and places thymine in the template strand. A two hydrogen bond, non-aromatic nucelic acid, thymine, has replaced the original three hydrogen bond, non-aromatic nucleic acid, guanine. The gene is now mutated because it no longer has the correct nucleic acid sequence.

Nevertheless, gene mutation does not necessarily lead to cancer. It's all luck of the draw. For example, the segment in the template strand of DNA containing the adenine-guanine-guanine (AGG) sequence codes for thymine-cytosine-cytosine (TCC) in the replicate strand. During cell division, polymerase reads TCC of the replicate strand to synthesize the original DNA template AGG. This triplicate in the normal gene of DNA codes for uracil-cytosine-cytosine (UCC) in the synthesis of RNA. The UCC sequence in RNA codes for the amino acid serine in the protein being formed, which is often an enzyme.

If the first guanine is alkylated at the O^6- position, polymerase reads a non-aromatic, two hydrogen bond nucleic acid, or thymine. Polymerase will place thymine's complementary nucleic acid, adenine, in the replicate to form the triplet TAC. This time when the cell divides, polymerase reads this triplet, TAC, in the replicate strand and synthesizes a new triplet ATG, instead of AGG, to place in the mating strand. The gene is now mutated. The triplet ATG in DNA codes for UAC in RNA synthesis, which in turn codes for the amino acid tyrosine. When the protein is synthesized, tyrosine replaces serine. If this amino acid substitution is at a key site in the catalytic pocket (the place in the enzyme that alters chemicals, endogenous and foreign), enzymatic activity will be lost. If this enzyme is required to deactivate the endogenous chemical messenger that stimulates cell division in a given cell type, the stop switch is permanently lost, resulting in cancer.

If the second guanine is alkylated in the AGG triplet, polymerase will read thymine and synthesize the TCA triplet in the replicate strand. When the original DNA template is reproduced, the triplet will be AGT. This triplet codes for UCA in RNA, which also codes for serine. Although in this case a mutation occurred in the DNA sequence, no change in protein function will be observed. No cancer will result. The person got lucky.

If both guanines are alkylated in the AGG template triplet, polymerase will read ATT and synthesize TAA in the replicate strand. It will then synthesize ATT, instead of AGG, in the new template strand. ATT in DNA codes for UAA in RNA synthesis, which is the code for termination of protein synthesis. The length of the protein will be determined by the location of the original, mutated AGG triplet. An incomplete, nonfunctional protein is produced. If it is the messenger deactivating enzyme, then cancer results.

These are the three combinations for alkylation of this AGG triplet. Alkylation of the second guanine causes a mutation that will not be observed because the same amino acid is placed in the protein chain. Alkylation of the first guanine causes tyrosine to replace serine in the protein. If this protein is the messenger deactivating enzyme, cancer results. Alkylation of both guanines leads to the termination of protein synthesis. Again, if this protein is the deactivating enzyme, cancer forms. The odds are 2 out of 3 that you will get cancer if enough DNA is alkylated at the hydrogen bonding oxygen. And then only if the affected protein is the deactivating enzyme for the endogenous messenger .

Alkylation of any of the four nucleic acids at the N^3- and N^7- positions does not impair the ability of polymerase to read the number of hydrogen bonds and aromaticity of the nucleic acid; therefore, the correct nucleic acid is added to the DNA sequence. Cancer will not result.

Guanine is in other triplets that code for amino acids, but this AGG triplet is one of the best examples to illustrate various consequences of guanine alkylation, from no damage to the cell to the formation of cancer, or even cell death. Which of these consequences results is a question of numbers, of how many guanines are alkylated, which guanines are alkylated, and the luck of the draw.

The above example illustrates how alkylation of only one guanine at the O^6- position could lead to a substitution of only one nucleic acid that would cause the substitution of only one amino acid which would destroy the function of the enzyme. In an analogous series of steps, alkylation of thymine at the O^4- position leads to miscoding of DNA that is ultimately observed as changes in protein function. When the number of alkylated guanine and thymine nucleic acids increase such that the de-alkylase enzyme, which repairs DNA by removing the added alkyl moiety, gets overwhelmed, the chances of significant alterations in genes (mutations) also increase to produce cancer in a given single cell.

Alkylation of DNA is just one process that can lead to genetic mutations. Some chemical carcinogens intercalate (are sandwiched) between the nucleic acids, like the antihistamine methapyrilene. This causes a shift in the genetic code. If this damage is not repaired, then as polymerase comes across the carcinogen, it can substitute any of the four nucleic acids (adenine, cytosine, guanine, or thymine), or none, in this location. The same process of miscoded gene replication described for alkylated DNA will occur. A mutation may or may not result, depending on the luck of the draw. If this mutation is in the gene that codes for the deactivating enzyme, then cancer results. There is more than one way to produce the same net effect of altering the genetic code and causing the formation of cancer.

Carcinogenesis from Radiation

Carcinogenesis resulting from radiation exposure is a variation of the mechanism with chemicals. It truly is based on the principle of "luck of the draw".

Due to their relatively large size and short distance of travel, it is unlikely the alpha particles enter the cell wall to cause damage to genes. This form of radiation probably does not cause cancer, although it is not necessarily harmless.

Beta particles can readily penetrate the cell wall without hitting biomolecules due to their small size and longer distance of travel. A single beta particle will cause little, if any, damage. But the more exposure (increased number of particles) to beta radiation, the greater the chance of one or more hitting and damaging the gene coding for the deactivating enzyme. Because they travel in a linear path and are relatively small, beta particles can readily pass through one organ without hitting any biomolecules only to cause damage in the next organ. They may even pass out of the body without hitting any biomolecules.

X-rays and ultraviolet (UV) light, which are forms of gamma rays, usually pass through the body without hitting biomolecules. Nevertheless, as the exposure level increases, so does the chance that they will hit biomolecules. The tissue in the front line of exposure to this type of radiation is the most likely to get damaged. Skin is the first tissue to receive UV rays, hence, cancer forms there most often. X-ray machines bombard a focused area of the body, the breast or lung, for instance, resulting in tumors of these organs. Because gamma rays also travel in a linear path, tumors form in whichever organs they happen to hit. UV light causes breaks, called nicks, in chemical bonds in DNA. There is another group of enzymes in the nucleus, called nucleases, that patrol the DNA molecule to correct this damage. Their function is to cut out the nicked DNA fragment. Polymerase then synthesizes a new, correct fragment to replace the previously damaged location. If this damage is repaired before the cell divides, then no adverse effects result. If the nick is not repaired in time, and it is in the gene that codes for the deactivating enzyme, cancer results.

Unlike chemicals, radiation can not target organs. However, it can produce the same type of key damage to the genetic code. The formation of cancer in any cell type derives simply from the statistical probability, which can not be calculated, of the radiation hitting the gene coding for the deactivating enzyme, with the organs closest to the source of radiation having the greatest chance of becoming cancerous.

The Hegedus Theory of carcinogenesis explains the mechanism of carcinogenesis, but it can not be fully functional unless it explains the dichotomies that have been well documented in cancer research. A detailed discussion of the vast array of literature and references along with the presentation of procedures I used and the results obtained is beyond the scope of this book. Some of this data is included in this volume. A more complete discussion of these subjects will be dealt with in a companion volume titled "Carcinogenesis: The Hegedus Theory", which will be available soon.

CHAPTER **21**

RESOLVING THE DICHOTOMIES

The principles outlined by the Hegedus Theory can be applied to explain the dichotomies concerning cancer reported in the literature, and to answer some of the most frequently asked questions.

Alkylation of DNA

As noted earlier, there have been four reported combinations of low and high concentrations of DNA adducts, and tumors or no tumors. These are now readily explained.

1. High concentrations of alkylated DNA lead to the formation of tumors when the damaged gene coding for the deactivating enzyme can not be repaired before the cell divides; the mutation is passed on to the progeny cells.

2. On the other hand, experimentally measured high concentrations of alkylated DNA may not lead to tumors for several reasons. One, the gene coding for the deactivating enzyme has not been damaged. Two, while the DNA molecule may be extensively damaged, the deactivating enzyme remains intact to keep the concentration of messenger from reaching the second critical

concentration. When dosing by the carcinogen stops, the cell has enough time to repair the damage to all of the genes. Three, the entire organ is homogenized in the experimental procedure with the amount of alkylation reflecting the weighted average of the cell types. What appears to be a high concentration of alkylated DNA is actually the sum of many cells with low concentrations of alkylated DNA. The carcinogen may not have targeted the cell type in that organ that forms cancer. The gene for the deactivating enzyme is not altered and no cancer results. And four, most of the DNA alkylated adducts reported are at the N^3- and N^7- positions which do not lead to the miscoding of genes.

3. A low concentration of DNA alkylated adducts and the formation of tumors means that the carcinogen has a high affinity for the receptor and the deactivating enzyme. Due to the selective uptake process by the cell type that gives rise to the tumor, a high concentration of the carcinogen is present, with low concentrations in non-target cells of the organ. The deactivating enzyme rapidly produces reactive metabolites of the carcinogen. The metabolites inactivate the enzyme and alkylate its gene. The messenger reaches the second critical concentration and the cell divides to produce two cancer cell progenies that eventually give rise to a tumor. Because a low concentration of the carcinogen exists throughout the organ, the amount of alkylated DNA adducts is low, but not in the target cells.

4. Low concentrations of alkylated DNA adducts and no tumors simply indicates that either very little carcinogen entered the organ, or that it passively enters the cell and is metabolized rapidly to non-reactive metabolites. Damaged DNA is repaired before the cell divides.

DNA Fragments and Protein Content

Molecular biologists have reported a range of differences between DNA fragmentation patterns of normal cells and cancer cells. No change to many changes have been reported. Biochemists have reported similar patterns in the protein content of normal and cancer cells.

Both researchers use gel electrophoresis to separate DNA fragments and proteins according to size and the charge to mass ratio. This separation technique takes advantage of gross differences in size and nucleic acid sequences in DNA fragments and amino acid sequences in proteins. Minute changes in the structures of these huge biomolecules can not be detected with this technique.

If one nucleic acid substitution occurred that leads to the miscoding of only one amino acid in the catalytic pocket of the deactivating enzyme, no change in the fragmentation pattern of the DNA will be observed. The slight change in the amino acid sequence also will go undetected. Nevertheless, cancer forms in the organ. The molecular biologist and the biochemist will report no changes. It is only when a significant alteration in the gene occurs which leads to a significant alteration in the structure of the protein that these changes are detected using gel electrophoresis.

It is possible to detect minute changes in the nucleic acid and the amino acid sequences, but it is not practical. All that needs to be done is to isolate the gene and the enzyme, and then sequence each one. To give you an idea of how monumental this task is, researchers have received Nobel Prizes for the isolation and sequencing of one gene, or for the isolation and sequencing of one protein. If you recall, it costs about $150 million to locate a gene and about $2 to $4 million dollars to isolate a new enzyme. In addition, it takes from 3 to 6 years to isolate a gene, and about 2 years to isolate an enzyme. An additional 3 to 6 years is required to sequence the gene, and about 2 years to sequence a protein. You can understand why detection of minute changes in the sequences of genes and proteins is not routinely attempted. In the end, all that would be accomplished anyway is the verification of the obvious - that a substitution in the nucleic acids of the gene caused by a carcinogen led to a miscoding for the deactivating enzyme such that it can no longer deactivate the endogenous chemical messenger which stimulates cell division leading to the initiation of carcinogenesis. It's much ado about nothing.

Lung Hyperplasia

Researchers have reported that both hypoxia (below normal oxygen concentration) and hyperoxia (above normal oxygen concentration) produce hyperplasia (a proliferation of normal cells which recedes with time) in the lungs of all animals, including rats, hamsters, and humans. Hypoxia in humans is observed in miners and heavy cigarette smokers. Hyperoxia is experienced by pilots and astronauts. This hyperplasia comes from a proliferation of neuroendocrine cells in the lung, which are rare in the normal adult lung. This cell type is common in mammals soon after birth, but then disappears with no apparent ill effects. Hyperplasia of neuroendocrine cells can be a potentially lethal condition because this cell type does not facilitate the proper exchange of oxygen and carbon dioxide in the lung. However, this cell type is the only major producer of the necessary neurotransmitter serotonin other than the brain. It is extrememly important to understand the pharmacological and biochemical characteristics of the neuroendocrine cell because it gives rise to more than 70% of the lung cancer cases in patients who smoke.

Here we are faced with another dichotomy in that both low and high oxygen concentrations produce the same clinical condition of hyperplasia in the lung. This can be explained. Neuroendocrine cells manufacture and secrete into the bloodstream the neurotransmitter serotonin, and selectively uptake it for metabolic deactivation. The serotonin, which has the official chemical name of 5-hydroxytryptamine (5-HT), is deactivated by the enzyme monoamine oxidase (MAO) which removes an amine (NH_2) from the ethylamine moiety ($-CH_2CH_2NH_2$) of serotonin and replaces it with an oxygen atom to form an aldehyde moiety ($-CH_2CHO$). The resulting metabolite, 5-hydroxyindoleacetaldehyde (5-HIA) is passed out of the cell and eliminated in the urine under normal circumstances.

Notice that in this pathway, oxygen is required to deactivate serotonin. In hypoxic conditions, not enough oxygen is available for MAO to metabolize serotonin to 5-HIA. Consequently, the intracellular concentration of serotonin increases to reach the first critical concentration. But in this case, more enzyme can not metabolize the messenger due to the low oxygen concentration. The second critical concentration of serotonin is quickly reached, causing cellular division. With thousand of cells undergoing this process, it does not take long for hyperplasia to form.

Hyperplasia resulting from hyperoxic (high oxygen concentration) conditions is a more complex process. There certainly is plenty of oxygen to aid MAO in metabolizing serotonin, but the product of the enzymatic reaction, 5-HIA, can bind with MAO and inhibit the further metabolism of serotonin. This is called feedback inhibition. The cell still manufactures, secretes and uptakes serotonin at the same rate, but it is metabolized at a reduced rate because 5-HIA ties up the MAO. As a result, the intracellular concentration of serotonin builds up through both critical concentrations to stimulate cell division. Again, with thousands of neuroendocrine cells dividing, hyperplasia in the lung results. The rate at which this condition occurs is slower than with hypoxia because more cells (hence enzyme capacity) metabolize serotonin at a faster rate to keep the messenger from temporarily entering the nucleus. However, as the number of neuroendocrine cells increase, so does the amount of serotonin, which causes this defense mechanism to be overridden, and the cell divides.

Onset of Carcinogenesis

What appears as the sudden onset of cancer is really the culmination of a long process that reduces down to a question of numbers. In the beginning of carcinogenesis, literally only one cell in the organ may be cancerous. As you learned, the growth rate of cells is dependent on the concentration of chemical messenger. In the early stages of cancer, there is a low number of cells and a low concentration of messenger. In addition, the tumor needs new vascularization (blood vessels) to feed it and to eliminate wastes like normal cells. This process takes time. In fact, many cases of spontaneous remissions are probably due to the tumor outgrowing its supply of blood vessels. Consequently, the growth of the tumor is initially slow. Also, it is impossible to detect single cancer cells using the standard techniques of x-ray and biopsy. In the initial phases of cancer, the tumor cells are not tightly packed and can not be distinquished from normal tissue. The extremely small tumor escapes detection. Also, the patient does not experience any discomfort, so there is no reason to see a physician. As the tumor size increases, effects are felt by the patient. For example, a breast tumor is detected because the number of cells in the tumor is so high that they pack closely together. This tight packing of cells shows up on film as a large light spot. Sometimes patients complain of pain in a region of the body and visit the physician who may perform a biopsy which tests positive for cancer.

Before these effects are felt or observed, the tumor grows. Cancer cells are not immortal. They wear out and die like normal cells. In the tumor, there is a constant dynamic equilibrium between the rates of cell division and death, with the rate of division far exceeding the rate of death. Cancer cells multiply at a constant rate and they die at a constant rate that is very much lower than the rate of division. This is another reason the initial rate of cancer growth is slow. The normal cell counterparts in the organ selectively uptake and deactivate much of the messenger to decrease its concentration. These dynamics can go on for years without discomfort to the patient while the tumor escapes detection. A patient can be exposed only one time to a high enough dose of a carcinogen to initiate carcinogenesis, but the tumor may appear 20 years later. Remember, cancer cells are not toxic to the body. A patient only feels pain or discomfort when, for example, the tumor is pressing on a nerve. Or in the case of leukemia, the high number of white blood cells displaces the red cells such that proper oxygen and carbon dioxide exchange can not occur, and the patient always feels tired. These adverse effects are not due to a poison excreted by cancer cells.

Another reason there appears to be a time lag in the detection of a tumor is the nature of carcinogens. In general, carcinogens are not very toxic. A person can be exposed to relatively high doses without feeling any adverse effects. It is possible, and likely, that a person is exposed to high concentrations of a carcinogen for long periods of time and not know it. Nitrosamines found in tobacco products and industrial pollutants are excellent examples of relatively nontoxic carcinogens we are exposed to for long periods of time without noticing adverse effects. In contrast, carbon monoxide is not a carcinogen, but it is very toxic. Even though we can not see, taste, smell, or feel this poisonous gas, we certainly quickly experience its effects of nausea, lightheadedness, and shortness of breath. When experiencing these symptoms, we leave the source of carbon monoxide to end our exposure.

As the tumor grows, more cells synthesize the messenger, increasing its concentration. More messenger stimulates a faster growth rate of the cells. More cells increase the concentration of the messenger, and more cells divide. This snowballing effect in growth rate continues until a tumor appears. In addition, I have demonstrated that continued exposure to carcinogens indirectly accelerates the growth rate of cancer cells. Many people, especially smokers, continue to expose themselves to the carcinogen that caused their cancer, which accelerates the growth of their tumor.

It can seem that a woman did everything right to avoid breast cancer by getting mammograms at appropriate intervals, and then suddenly a tumor appears. This process was not instant. It was years in developing, but only recently could the tumor be detected in that patient. No oncogene made a sudden burst for glory by coming out of the closet to express itself.

Alleged Oncogenes

Sometimes molecular biologists actually see differences in the location of a DNA fragment band between the normal and cancer cell. Due to several alterations in the nucleic acid sequence in the cancer cell, the gene coding for the deactivaing enzyme probably migrates a different length in the gel than the normal gene. Molecular biologists are detecting the genes for the various altered deactivating enzymes and calling them oncogenes. These altered genes code for nothing. It is not surprising that when they put these useless DNA fragments into normal cells with the desire to initiate carcinogenesis, nothing happens.

Time Bomb Carcinogenesis

Throughout this book, I have repeatedly stressed that no one is genetically programmed to get cancer. As with all rules, there are exceptions. The retinoblastoma of infants is one. Some types of childhood leukemia are others, although I am not certain what the endogenous chemical messenger is particular for these types of cancer. However, I would place my money on either histamine or serotonin. If this is true, then the enzyme would be another form of MAO than the one specific for serotonin. Certainly, these children were born without the enzyme to deactivate the messenger that stimulates the division of white blood cells.

Another type of cancer forms in pubescent girls born to mothers who took the fertility drug diethylstilbestrol. At the age of about 13, these girls die of uterine cancer. Because progesterone stimulates the proliferation of endometrial cells in the uterus, these girls were probably born without the gene that codes for the diaphorase enzyme that deactivates this hormone. This leads to fatal uterine cancer. Cancer of the uterus does not show up earlier because the uterus does not start to develop until puberty, which usually is about the age of 13. At this age, the uterus begins to grow but it can not stop in these girls. Uterine cancer results.

None of the children with these cancers lives beyond puberty; therefore, they never can produce offspring who will inheret these genetic defects that cause cancer. In accordance with the Hegedus Theory, the same mechanism that is in operation in the adult occured in the fetus.

Potency of Carcinogens

Some of the reasons for differential potencies of carcinogens were discussed in a previous section. The main reason is due to the interaction between a carcinogen and a receptor. Ariens developed his Receptor Occupancy Theory to explain the differential potencies of drugs in a series of congeners (drugs that have the same basic structure that has been slightly altered in each compound). The potency of the drug is proportional to its affinity for the receptor. The higher the affinity, the more potent the drug. Drugs are chemicals. Most carcinogens are chemicals. The Ariens theory is applicable to carcinogenesis.

According to Ariens, the effect is due to a drug occupying a receptor such that a drug-receptor complex forms. For chemical carcinogens, the effect is carcinogenesis, the chemical carcinogen substitutes for the drug to form a carcinogen-receptor complex. Two assumptions are made with this theory:

1. No covalent, permanent, bonds are formed between the carcinogen and the receptor.

2. No other compound occupies the receptor.

Generation of a stimulus (carcinogen metabolism or carcinogenesis) is directly proportional to the concentration of the carcinogen-receptor complex. The maximum chance for cancer to form is when all of the receptors are occupied by the carcinogen. In the simplest case, there is a linear relationship between effect and and the stimulus. However, a number of interactions affect the relationship between carcinogen and the receptor, which complicates this relationship between effect and stimulus.

One of these factors is the intrinsic affinity the carcinogen has for the receptor to form the carcinogen-receptor complex. The carcinogen with the highest intrinsic affinity will produce the greatest effect. Conversely, carcinogens with low intrinsic affinity for the receptor will produce minimal, if any, effects.

Another factor is how the carcinogen partitions throughout the body. The partitioning characteristic of the carcinogen is dependent upon the degree of lipophilicity (fat soluble) versus hydrophilicity (water soluble) characteristics inherent in the molecule. As lipophilicity increases, affinity for the receptor increases to a maximum point, and then decreases. Pharmacological receptors also have a hydrophilic region. As the hydrophilicity of a compound increases, so does its affinity for this part of the receptor until a maximum is reached, then the affinity decreases. Optimum affinity for a receptor is achieved when a proper balance exists between the lipophilic and hydrophilic characteristics of a chemical. Potency of the carcinogen and targeting of the receptor is due to these two intrinsic characteristics of the molecule.

Pharmacological receptors are designed to detect very small differences in the structures of chemicals, although some tolerances are made. For example, the hormone estradiol has only one more hydrogen atom located on the hydroxyl group of the D ring than estrone, which has a carbonyl group at this location. While estradiol and estrone have almost the exact same structure, estradiol is 12 times more potent than estrone, which means it has 12 times more affinity for the estrogen receptor. Just as pharmacological receptors make distinctions between very subtle differences in the structures of endogenous chemicals, like hormones, they

make such distinctions with foreign chemic
carcinogens as well.

Metastasis

Taber's defines metastasis as, "Movement of bacteria or body cells (especially cancer cells) from one part of the body to another." I do not believe this definition is accurate for cancer cells. I know that a patient can have a large primary tumor and several smaller secondary tumors. But the secondary tumors did not originate from the primary tumor. Instead, the same mechanism of carcinogenesis that formed the primary tumor also caused the secondary tumors. The organ that has the primary tumor contains most of the cell types which selectively uptake the carcinogen via a pharmacological receptor; therefore, these cells are the most likely to become cancerous.

The same type of receptor can be located in other organs, with a fewer number of cells. For example, serotonergic receptors are located in the respiratory system, cardiovascular system, smooth muscles of the alimentary tract and blood vessels, exocrine glands of the gastrointestinal tract, nerve endings including a cluster of nerves called autonomic ganglia, adrenal glands, and blood platelets. Estrogen receptors are found in the sexual organs, breast, skeleton, fat, liver, and skin. Any given one endogenous chemical messenger has receptors in several organs. The same cell type that synthesizes the messenger in the primary organ is present in the other organs as well.

Each metastatic tumor is actually a localized tumor formed from the cells of that organ. Because the primary organ uptakes most of the carcinogen, the secondary organs receive very little exposure. Whatever amount they receive is harmlessly metabolized. Once the primary tumor forms, it can not metabolize either the messenger or the carcinogen. Consequently, the concentrations of both chemicals increase in the body. Eventually, the same deactivating enzyme in the secondary organs as in the primary organ is overwhelmed and its gene is destroyed due to the increased amount of carcinogen. This destruction coupled with the increased concentration of messenger received by the cells from the primary tumor initiates carcinogenesis in the secondary organs. Because carcinogenesis is dependent on the concentrations of carcinogen and messenger, it takes a longer period of time for the metastatic tumors to appear.

One of the main sources of error in conducting experiments with tissue culture of cancer cells is that they clump together. Even when they are shaken in an attempt to suspend them as individual cells, many adhere to each other. The cells on the border of the clump uptake the chemical, but the inner cells are not exposed. The measurements of metabolism are an average of those cells which uptake the chemical and those not exposed to the chemical. Because each clump is a different size, each measurement is different. After the metabolism part of the experiment is completed, the vials with the cells are placed into the refrigerator overnight to be kept for protein content determination the next day. In the morning when the vials are removed, one large clump of all the cells is in each vial. Even in the conditions of tissue culture, cancer cells adhere to each other. If cancer cells in tissue culture have such a strong adhesion for each other, then the adhesion would be stronger in the body. Cancer cells do not slough off and enter the blood stream to form tumors in other parts of the body.

A number of researchers have examined the effects of anti-cancer drugs on primary and secondary tumors. They took cross section slices of the primary tumor from the periphery to its center. No difference in response to the drug was observed by any of these cross sections. They performed the same experiment on secondary tumors, and obtained the same results. However, when the responses of the primary tumor were compared to the responses of the secondary tumor, they were different. Moreover, some researchers have reported different responses among the secondary tumors. If all of the tumors originated from the exact same cell, then the responses to the drug should have been the same regardless of which organ the tumor was taken from. This differential response by secondary tumors to a drug shows that metastatic tumors do not originate from the primary tumor. The data does not support the decree that metastatic tumors come from the primary tumor.

Carcinogens Are Not Necessarily Mutagens

Many carcinogens have not produced any alteration in the genetic code of bacteria as is currently measured. A change may have occurred but it is not detected by methods routinely used. All microorganisms have the capacity to uptake huge chunks of DNA and incorporate them into their genetic code. Often, the microorganism actually transcribes these DNA fragments into some form of protein. It is the mechanism whereby they constantly mutate for survival. But there is no reason to assume that the proteins produced will either be useful or harmful to the microorganism. For example, human insulin that is produced by bacteria as a result of genetic engineering is neither useful nor harmful to the microorganism. Its enzymes, in robotic fashion, just read the genetic sequences and produce the proteins they code for. While many chemicals can alter the genetic code, they do not affect the normal operation of the microorganism. This is the same mechanism described by the alkylation of guanine that may or may not lead to cancer.

However, the most likely explanation is the simplest. The chemical never was able to enter the cell. Remember, cell walls, or membranes, are designed to keep out undesirable chemicals. If the chemical is unable to enter the cell, it can not produce an effect. Consequently, no mutation will occur.

Benign Versus Malignant Tumors

There is no such entity as a benign tumor. If the growth is cancerous, it will continue to grow at a rate dependent on the concentration of the endogenous chemical messenger.

Hyperplasias which most often are misdiagnosed as a benign tumor are a slow growing mass which also is dependent on the concentration of the endogenous chemical messenger. However, when this concentration recedes, the growth rate slows until it eventually stops, the tissue recedes. It should be understood that this process takes time. During the relatively rapid growth phase of the hyperplasia, it can, and often is, exposed to any of a number of carcinogens. According to the Hegedus Theory, all rapidly growing cell types are the most likely to transform into a cancer because their genetic codes can be altered for the deactivating enzyme before there is a chance to correct it. While almost all chemical carcinogens are selectively uptaken by a cell type, all of them enter also by the mechanism of passive diffusion into other cell types to damage the DNA molecule. If this damage occurs during the state of cell division, it is unlikely that the repair enzymes can repair the damage to the gene coding for the deactivating enzyme. One carcinogen may have produced hyperplasia, but it is possible that another one caused the transformation inadvertently to cancer.

Using the light microscope, it is imposible to ascertain the distinction between cancer and hyperplasic cells. As described previously, the isolation of the NCI-H727 lung neuroendocrine cell line was due to the misdiagnosis by light microscopy of a malignant tumor that was actually "benign".

If your physician tells you that you have a so-called benign tumor, have it removed immediately. No harm can come to you. If it is malignant, then you caught it in its earliest stages, which will increase your chances of survival. Even if your tumor is diagnosed as malignant and inoperable, have as much of the tumor removed so that the growth rate is substantially reduced. You now know that this rate is dependent on the concentration of endogenous chemical messenger. Reduce the number of cells that produce this messenger and you reduce the tumor growth rate, BUT, only for a period of time. However, you will prolong your life by a significant amount.

Asbestos

Asbestos is not a carcinogen. It is a victim of the witch hunts discussed in a previous chapter. Not only did bimbos relentlessly seek out this relatively harmless mineral with their "statistical proofs" and make huge sums of money from their bloodthirsty hunts (and had their voracious egos fed), but substantially lesser minds made perhaps even more money allegedly "cleaning up" this non-reactive material.

All compounds are toxic in extreme exposure, or dose. The necessary element for our lives, oxygen, kills at a high enough dose. Water, without drowning, can kill. The men who developed lung cancer or emphysema from asbestos exposure spent 10 to 12 hours per day for typically 7 days a week during World War II to line the insides of battle ships for the United States Navy. Under these conditions, they were exposed to very high levels of asbestos dust, which is a lung irritant at these concentrations. In addition, most of them smoked cigarettes or cigars before, during, and after working around asbestos.

Asbestos can form microscopic needlelike fibers that can penetrate and destroy cells of every type. Because these fibers were in the air, they readily were inhaled and penetrated lung cells.

According to the Hegedus Theory, damaged cells release endogenous chemical messengers that are selectively uptaken by remaining cells of the type that produce it. Therefore, when these lung cells were damaged, they released a messenger, serotonin, to accelerate the growth rate of the other identical cell types in the lung that were not damaged. This increased the intracellular concentration of the messenger and the cells divided. If there was any damage at this time to the gene coding for the deactivating enzyme, then lung cancer resulted. But the genetic damage came from a secondary exposure to a chemical carcinogen, most likely from a tobacco product. Even if the worker did not smoke, his co-workers did in an enclosed environment, much like the driver who smokes on his way to work sitting in his car. Non-smoker and smoker alike inadvertently received bolus doses by some of the most volatile and potent chemical lung carcinogens, diethyl and dimethylnitrosamines. It is the damage to lung cells due to the microscopic needlelike particles coupled with the exposure to potent lung carcinogens that led to the formation of cancer in this organ. The amount of asbestos required to cause this damage is very much higher than the Environmental Protection Agency (EPA) guidelines. Guidelines for exposure to asbestos are set so low simply because a few contractors are making millions of dollars removing it, while the egos at the EPA are fed daily. There is no scientific rationale for these limits.

Mothers, don't have your babies eat asbestos, but don't panic if they are around this material. There is no justified reason for it to be removed from schools and public buildings other than when it decomposes and loses its insulating properties.

Silicone Breast Implants

Silicone is a relatively inert chemical; therefore, it is not expecially toxic or carcinogenic. However, if the implant should leak, there is the possibility for silicone to become oxidized to form silicon dioxide. The common names for this chemical are sand and quartz. Should this reaction take place, women can literally be having their breasts sandpapered from the inside. Scar tissue will form which is harder than normal breast tissue. Damaged breast tissue divides more rapidly than normal, leading to an increased chance of a carcinogen causing the destruction of the deactivating enzyme for the breast's endogenous chemical messenger. While silicone is not a carcinogen, it can contribute to the formation of breast cancer.

Carcinogenesis and You

Theory is only as good as its ability to describe and to predict reality. This chapter has shown how the Hegedus Theory can explain and, in fact, predict the necessary dichotomies reported in the literature. The next chapter shows how the Hegedus Theory can explain and predict the mechanisms of particular cancers using as examples lung, breast, and an especially insidious form of infantile eye cancer. Emphysema, although not cancer, is discussed as well. The Hegedus Theory has unravelled the mystery of emphysema's similarity in development to lung cancer while at the same time precluding carcinogenesis in this organ.

CHAPTER 22

CANCERS

The principles of the Hegedus Theory can be used to explain how and why particular chemicals target specific cell types and cause cancer.This is true of any carcinogen and every cancer. What follows is an explanation for three of the most vile cancers - lung, breast, and a form of eye cancer that afflicts newborns. An explanation for the onset of emphysema is included because this non-cancerous but equally malignant disease has puzzled scientists for decades. The Hegedus Theory explains the similarity in the development of emphysema and lung cancer, and why emphysema sufferers do not typically develop lung cancer.

Lung Cancer
According to the Hegedus Theory, the following series of events leads to lung cancer in humans. Cigarette smokers inhale large amounts of carbon monoxide from tobacco smoke, leading to a decreased level of oxygen in the lung, or hypoxia. This hypoxic condition stimulates the proliferation of lung neuroendocrine cells because MAO does not have enough oxygen to deactivate serotonin, and the second critical concentration of the

messenger is reached. In addition, the diethylnitrosamine (DEN) inhaled with the smoke is selectively uptaken by the serotonergic receptors on the membranes of the increasing numbers of neuroendocrine cells in the lung. The DEN is metabolized by monoamine oxidase (MAO) to produce reactive metabolites that alkylate DNA and that in turn inactivate MAO which already is working at a reduced capacity due to the decreased concentration of oxygen. These two factors combine to produce an increased concentration of the endogenous chemical messenger, serotonin, which reaches the first critical concentration, stimulating the synthesis of more MAO. This defense mechanism is overwhelmed rapidly due to the low amount of oxygen and the relatively high concentration of DEN in the lung. Consequently, the second critical concentration of serotonin is reached and the cells divide in an attempt to fight the onslaught of the carcinogen.

As long as the DEN concentration remains relatively low in the lungs, even though higher than in non-smokers, MAO can metabolize DEN and serotonin. Alkylated DNA adducts that form are removed by the de-alkylase enzyme. No cancer forms.

In the first few years of tobacco use, the smoker's lungs can readily handle this metabolic overload. Gradually, the number of neuroendocrine cells increases, and the oxygen concentration in the lung decreases due to tars deposited on the linings, i.e., epithelium, of the lungs. Hyperplasia results which is still not life threatening, and no ill effects are felt yet by the smoker. With the increased number of neuroendocrine cells that selectively uptake the carcinogen, the chances of causing irreparable damage to the genetic code, especially to the gene coding for the deactivating enzyme, sharply increase. With cells that are in a state of rapid division, the chances are substantially increased for mistakes to occur in the synthesis of genes, and thereby, cancer to form. This process can take years.

Let us look at the pattern of tobacco use among smokers as an explanation for why carcinogenesis can take years to appear. The typical heavy smoker has a cigarette every 15 to 20 minutes while in a social environment that allows smoking. Under these circumstances, the dose of DEN is relatively low, 3 to 5 cigarettes per hour. During non-smoking periods, there is no additional exposure to the carcinogen. The cell has time for the de-alkylase enzyme to repair damage caused to the DNA by removing the alkyl moiety, and to synthesize more MAO if necessary. Also, because the enzyme is not occupied with the metabolism of the carcinogen, it can regulate the intracellular concentration of serotonin. This is not to imply that no smoker can get lung cancer smoking less than 3 to 5 cigarettes an hour. Remember, any exposure to a carcinogen sharply increases the chance of getting cancer.

When a smoker is alone, the number of cigarettes smoked increases. Often the smoker is in the car with the windows rolled up. In this environment, less oxygen containing air enters the car, and more volatile carcinogens, like DEN, remain inside the car. Now the smoker gets a bolus dose of carcinogen from rebreathing the carcinogen which is exhaled while inhaling it from the cigarette. This pattern of smoking goes on for years.

At this point, the smoker's lung becomes hyperplasic with neuroendocrine cells. There is more serotonin in the bloodstream to be uptaken simply because there are many more cells that synthesize and secrete it. DEN which ordinarily is metabolized by other cell types, like the Clara cells of the lung, or others in the liver, is now selectively uptaken in a large quantity by the growing number of neuroendocrine cells which are primed for rapid division. MAO metabolizes DEN to reactive metabolites that inactivate the enzyme and alkylate DNA at the O^6-guanine and O^4-thymine positions. The de-alkylase enzyme is overwhelmed by the number of alkylated DNA adducts, and it can not repair the damage in time. Meanwhile, MAO is inactivated by these metabolites which causes the intracellular concentration of serotonin to reach the second critical concentration. The cell divides. Wherever O^6-guanine and O^4-thymine are left, adenine or some other nucleic acid is substituted for the original nucleic acid. If this miscoding is at a location not necessary for proper cell function, i.e., recessive genes, and if the substitution does not change the amino acid sequence in the deactivating enzyme, no cancer results. The smoker is lucky, temporarily.

Eventually, a point of no return is reached such that the alkylation of DNA leads to a mutation of the gene coding for the deactivating enzyme, MAO in this case. This mutation is passed on to all of the progeny cells, which can no longer synthesize a functional deactivating enzyme. These cells still continue to synthesize and secrete serotonin at the same rate as their normal cell counterparts, but they have lost the ability to deactivate the messenger. Consequently it does not take long for the second critical intracellular concentration of the messenger serotonin to be reached, and the cell divides. This cycle of messenger synthesis and secretion, followed by selective uptake of it, inability of the enzyme to deactivate it which stimulates cell division continues until a tumor forms that is detected. The rate of tumor growth is dependent on the amount of messenger, serotonin, present. The higher the concentration, the faster the growth rate of the tumor. As the size of the tumor increases, so does its growth rate because more cells synthesize more messenger. Continued exposure to the carcinogen, like DEN, contributes to accelerate the growth rate of the tumor because it causes the same damage that produces carcinogenesis in additional normal cells.

Cancer does not form in all of the cells for several reasons, all based upon a question of numbers and luck of the draw. When a chemical, whether drug or carcinogen, enters the bloodstream, it weakly binds to a protein called albumin that carries it throughout the body. When the chemical comes into contact with a pharmacological receptor for which it has a greater affinity than for albumin, it detaches itself, a process called disassociation, and binds to the receptor. As you learned, the endogenous chemical messenger with which the chemical has the most structural features in common will determine the corresponding receptor for which it has greatest affinity. DEN is structurally similar to serotonin, so it has a high affinity for and is selectively uptaken by the serotonergic receptors.

Serotonin and DEN are in competition with each other for the serotonergic receptor. If the receptor is occupied by serotonin, DEN can not enter that particular cell, and it moves on in the bloodstream to the next one available. If all of the receptors are occupied, then DEN moves to the liver where it is metabolized at a reduced rate by MAO and the P-450 enzymes, into non-reactive metabolites that are eliminated in the urine; or DEN is excreted unchanged in the expelled air and urine. As the concentration of DEN increases, it competes more successfully for the serotonergic receptor such that more of it enters the cell and occupies MAO. Because at this point MAO does not metabolize serotonin, the second critical concentration that stimulates cell division is rapidly reached. During the time of exposure, there is a constant, dynamic competition between messenger and carcinogen at the pharmacological receptor and enzyme.

In general, cells closest to the point of exposure to a carcinogen have the highest probability of uptaking it. These are the cells with the highest probability of becoming cancerous. Cells further away from the point of exposure uptake less carcinogen simply because there is less available; therefore, they are less likely to transform into cancer cells. Remember, no exposure to a carcinogen means that no cancer can form. This is the mechanism that explains why one cell becomes cancerous while the cell literally beside it remains normal.

Emphysema

Emphysema is a common lung disease which very heavy smokers often get. It is defined in Taber's as, "A condition in which the alveoli of the lungs become distended or ruptured. Usually the result of an interference with expiration, or loss of elasticity of the lung." Distended means that there are open areas in the lung which ordinarily are not there. Loss of elasticity means that the lung can not expand normally as air is inhaled because scar tissue has replaced normal tissue. Breathing is more difficult. A tremendous amount of damage has occured in the lung such that cells are destroyed and not replaced, or replaced with nonfunctional cartilage. To date, no one has been able to explain how this damage occurs, but researchers know there is a link between very heavy cigarette smoking and emphysema without lung cancer.

Serendipity is an aptitude for making fortunate discoveries accidentally. Many medical discoveries were made this way. The accidental discovery of penicillin is one of them. Serendipity prevailed one day as I was working with human lung cancer cell lines. In preparation for cell growth kinetic experiments, I arbitrarily selected a concentration of DEN and serotonin to dose a neuroendocrine and a Clara cell line. I expected to accelerate the growth rate of the neuroendocrine cells with serotonin, but I did not expect the results observed with the Clara cells.

The day after initially seeding the flasks with both cell lines, I went to count them. The number of neuroendocrine cells dosed with serotonin was greater than the control cell group, which did not get dosed. The number of Clara cells in the flasks treated with serotonin was less than the control group. In the beginning of the experiment, the same number of cells was placed into each flask for both cell types. The missing Clara cells were not dead and floating. They were just gone. The next day, there were even fewer Clara cells in the treated flask, while the number of treated neuroendocrine cells increased more than the control. By day 5 of the experiment, there were no Clara cells in the treated flasks at all. During this time, the number of Clara cells in the control group continued to grow at a constant rate, as did both groups of neuroendocrine cells. It is highly improbable that someone was sneaking into the lab to steal treated Clara cells. Some mechanism was causing the disappearance of Clara cells treated with serotonin.

Very high doses of serotonin produced a toxic effect on neuroendocrine and Clara cells in a subsequent cell growth kinetic experiment. The number of cells in these two groups was very much lower than in the control and other treated groups. Electron microscopy showed that the enzymes in the lysosomes were released in the neuroendocrine cells after receiving a high dose of serotonin. These are the enzymes that destroy proteins by cutting them into pieces of amino acids. They continue to work until nothing is left of the cell. This is the reason they are contained in special sacs in the cell. The only logical explanation for the disappearance of the treated Clara cells is that serotonin stimulated the release of the enzymes in the lysosomes which then disintegrated the cells. Extremely high doses of serotonin had the same effect in the neuroendocrine cell as well.

This is the same situation present in emphysema, lung cells are destroyed and mysteriously disappear. In the patient's lung, a high concentration of serotonin exists which causes the release of lysosome enzymes in neuroendocrine and Clara cells. After these cells are destroyed, the enzymes continue to other cells and destroy them until these enzymes are stopped by destroying themselves. Wherever the localized concentration of serotonin is very high, lung cells disappear. This destruction does not happen all at once, but over a period of several years.

The mechanism of emphysema is related to the mechanism of carcinogenesis. Heavy smokers have a very low oxygen concentration in their lungs. MAO needs oxygen to metabolize serotonin and carcinogens. No oxygen is available for these reactions. The intracelluar concentration of serotonin rapidly accumulates to stimulate the release of the lysosome enzymes that destroy the neuroendocrine cell. Serotonin then enters the Clara cell where it stimulates the release of the lysosomal enzymes, destroying this cell type, also. The local tissue lesions are repaired by nonfunctional scar tissue. The net result is the formation of emphysema.

Lung cancer can not form because not enough oxygen is available in the lung for enzymes to metabolize carcinogens into the reactive metabolites that alkylate DNA, which in turn leads to the miscoding of the gene for the deactivating enzyme, and that inactivate the enzyme for the messenger. Cigarette smokers have the option of killing themselves with lung cancer or emphysema, depending on how many cigarettes they smoke per day.

Breast Cancer

It should be understood that there is no such thing as *an* estrogen. There are several estrogens, but the three most potent and prevalent are, in decreasing order of potency, estradiol which is about 12 times more potent than estrone which is 80 times more potent than estriol. These hormones are synthesized from androstenedione and secreted from the ovaries, placenta and adrenal glands, with the liver, fat, skeletal muscle and hair follicles as secondary sources in the form of estrone.

Estrogens, more specifically estradiol, produce an array of pharmacological effects throughout the body, but I want to limit the discussion to those effects on the breast. A more detailed description on the actions of estrogens can be found in any good text on human physiology or pharmacology.

In the breast, estradiol is responsible for the preparation of milk production by stimulating the stromal tissue and the ductile system. The lobes and alveoli are enlarged to a lesser extent. Estradiol also is responsible for producing the shape of the breast.

Estradiol is selectivley uptaken by target cell types in the breast, uterus, skeleton, and certain fatty tissues. Once inside the cell, it rapidly binds to a receptor. This messenger-receptor complex then migrates to the nucleus to stimulate production of RNA and certain proteins within a few minutes. A few hours later, DNA is duplicated leading to cell division, while the newly formed RNA initiates the production of proteins required for the specialized function of that cell type. This process is very similar to the one described for normal cell division, and the process whereby serotonin stimulates genes coding first for the production of proteins, eg. enzymes, followed by the division of neuroendocrine cells. This process is messenger concentration dependent.

If the intracellular concentration of estradiol were allowed to accumulate without regulation, the cells would be in a constant state of division. This steady, uncontroled rate of growth is the definition of cancer. There has to be an enzyme, similar to MAO for serotonin, that deactivates estradiol. The enzyme diaphorase, or oxido-reductase, removes a hydrogen atom from estradiol to yield estrone, which is less potent than estradiol. Estrone is further metabolized to estriol which is often eliminated from the body unchanged. However, diaphorase can also metabolize estrone back to estradiol should the concentration of the latter fall below normal. This enzyme is the key for regulating the intracellular concentration of estradiol.

Progesterone is another hormone secreted by the ovary, specifically the corpus luteum, with a structure similar to the estrogens. Although this hormone also has numerous pharmacological actions throughout the body, in the breast it stimulates the proliferation of the lobules and alveoli to prepare them for milk production, but it does not actually stimulate the production of milk. Progesterone augments the effects of estradiol. The precursor for its synthesis also is androstenedione. Progesterone is deactivated rapidly in the liver to a number of metabolites, including pregnanediol. This metabolite is formed by the same diaphorase enzyme that deactivates estradiol. Before the luteal phase of the menstrual cycle (the ovum has not left the ovary), pregnanediol makes up about 12 to 15% of progesterone urinary metabolites. During the luteal phase or pregnancy, 25 to 30% of the urinary metabolites consist of pregnanediol. This increase indicates that another organ is metabolizing progesterone, probably certain cells in the breast. The breast can not continue to grow indefinitely, so the messenger is deactivated in this organ as a means to regulate growth.

Once inside the cell, progesterone binds to a receptor that migrates into the nucleus to stimulate the synthesis of messenger RNA coding for a number of proteins. In animals pretreated with estradiol, the amount of progesterone receptor is substantially increased.

The net result of increased concentrations of estradiol and progesterone is to stimulate the growth of the breasts in preparation for the female to produce milk that will feed her baby. Breast enlargement during certain times of the menstrual cycle is a fact, but in the next phase of the cycle the breasts return to their normal size. Typically, this enlargement is relatively minor. However, it is certain that estradiol is the endogenous chemical messenger that stimulates the growth of one cell type in the breast, while progesterone stimulates another. Logically, if the deactivating enzyme is destroyed, then breast cancer will result. The challenge now becomes to determine the carcinogen(s) that is (are) most likely to be selectively uptaken by the estrogen receptor, or progesterone receptor, leading to a deactivation of the enzyme for estradiol. The role of the progesterone receptor and metabolic pathways must be considered in this mechanism because it too stimulates the growth of some of the same breast cell types.

Most chemical carcinogens induce the same pattern and incidence of tumors in both male and female animals of the same species. One carcinogen, bis-2-oxopropylnitrosamine (BOP), is interesting to study because it displays sexual dimorphism in the pattern and incidence of tumors induced in male and female Fisher 344 rats. Male rats develop thyroid, kidney and lung tumors. Female rats develop liver and lung tumors. The incidence of lung tumors in both sexes is relatively low compared to the 100% incidence of the two target organs in each sex: the thyroid for the male and the liver for the female.

In a series of experiments, BOP was given to six groups of male and female rats by mouth, gavage, and the metabolism of the carcinogen was measured by trapping exhaled $^{14}CO_2$ much like the experiments with the lung cancer cells. Group 1 was the normal female rat. Group 2 was the normal male rat. Group 3 was the female rat pretreated with estradiol. Group 4 was the male rat pretreated with estradiol. Group 5 was the male rat castrated from birth. And Group 6 was the male rat castrated from birth and pretreated with estradiol.

The group with the lowest rate of BOP metabolism was the normal male rats. Castrated male rats had a rate about equal to the female rats in the control group. In this group of male rats, testosterone, the male hormone, could not block the feminizing effects of estradiol because it could not be synthesized without the testicles. All of the groups pretreated with estradiol had significantly increased rates of BOP metabolism, with the group of pretreated, castrated male rats having the highest rate.

In a parallel experiment, these same six groups of rats were dosed with BOP in order to induce tumors. The three groups of male rats that were pretreated with estradiol, castrated, or both, developed the same pattern and incidence of tumors as the two female groups of rats. In the groups that received estradiol pretreatment, female and male, the tumors appeared in less time than in the other non-treated groups.

Clearly, estradiol changed both the rate of BOP metabolism and the pattern of tumor formation in rats. First, estradiol stimulates the synthesis of more enzyme which increases the rate of metabolism. Second, estradiol stimulates cell division which produces more enzymes that also increase the rate of metabolism. The pattern of tumor formation is dependent on the concentration of an endogenous chemical messenger, estradiol or progesterone. The challenge is to explain why this sexual dimorphism exists.

A logical start is to determine the tissue distribution of BOP. The highest concentrations should be in the target organs. More BOP should be found in the female rat liver than in the male, and more BOP should be found in the male thyroid than in the female thyroid. Little, if any, BOP should be found in non-target organs. This hypothesis is an application of the principles that if no cell comes into contact with a carcinogen, then it is impossible for it to become cancerous, and target cells selectively uptake the carcinogen.

When I analyzed for the tissue distribution of BOP in both sexes, the highest concentration of BOP was found in the target organs of both sexes, thyroid in the male rat, liver in the female rat, with an exception. In the female rat, an extremely high concentration of BOP's metabolites were found that could only be formed by diaphorase, rather than the original carcinogen. Others have reported high concentrations of alkylated DNA in target organs. These results are in keeping with the rate of progesterone metabolism in the liver. The high concentrations of metabolites show that an equally high concentration of BOP had to have been there initially for diaphorase to act on it. The selective uptake mechanism was actually verified by measuring the amount of metabolites.

I pretreated the female rats with the estrogen blocker drug, Tamoxifen, and examined the tissue distribution of BOP. The concentration of BOP was affected only in the target organs, indicating that the estrogen receptor is involved in the selective uptake of this carcinogen. BOP enters the target breast cell type via either the estrogen or progesterone receptor. Once inside the cell, it is metabolized, causing alkylation of the deactivating enzyme, diaphorase, and DNA such that the gene coding for this enzyme is destroyed. This allows the intracellular concentration of estradiol, or progesterone, to reach the second critical concentration that initiates cell division. The exact same series of events that led to carcinogenesis in the lung due to DEN is applicable for BOP, except that the deactivating enzyme and cell types in the affected organs are different.

BOP is a potent, volatile carcinogen that has been extracted from cosmetics, tobacco products, and cutting oils. While it is unlikely that many women use cutting oils, many do use cosmetics, and certainly virtually every woman is constantly exposed to tobacco products, especially the side-stream smoke not inhaled by the smoker. Even women who do not smoke are dosed with BOP. I believe that the steady increase in the incidence of breast cancer in women is due to their increased exposure to BOP from cosmetics, and the increasing number of women who use tobacco products.

Infantile Eye Cancer

There is a rare, but always fatal birth defect that is a cancer of the eye called retinoblastoma. Infants are born blind and usually retarded. It is a rapidly growing cancer that spreads in the brain, killing the child within about three years. These infants are lacking an enzyme called isomerase that deactivates the endogenous chemical messenger retinoic acid. β-Carotene, vitamin A, is the precursor for this endogenous chemical messenger. A gene that codes for isomerase has been mutated in these infants before birth such that retinoic acid can not be deactivated. Consequently, the regulatory switch for cell division is absent, and cancer of the eye results.

Retinoic acid binds to a specific cellular retinoic acid binding protein (CRABP) in the cytoplasm of the cell. This CRABP complex migrates to the nucleus to initiate a cascade of events, including cell differentiation (the special process to form the eye cell) and division. Eye cells that become cancerous use vitamin A as a precursor to synthesize an endogenous chemical messenger, retinoic acid, that stimultes cell growth. The enzyme, isomerase, that deactivates retinoic acid is nonfunctional in the cancerous eye cell, allowing the messenger to reach the second critical concentration that stimulates cell division. This is the same mechanism described for serotonin, estradiol, and progesterone. The normal eye cell is transformed into cancer because a key enzyme is lacking. The infant is born with this defect in the gene coding for isomerase; it is born with a cancer of the eye.

Several years ago, a molecular biologist received a Nobel Prize in medicine for claiming to have found the first of dozens of alleged oncogenes, which he named *ras*. According to his claim, retinoblastoma formed because this oncogene somehow got turned on, or expressed, in the infant. However, he could not offer a mechanism for how this alleged oncogene got turned on. It should be obvious by now, that it is *not* a gene that gets turned on that causes cancer. Cancer forms because a gene is *lacking* such that it *can not* be turned on. Putting in the useless step of introducing the concept of oncogenes does not explain carcinogenesis.

The formation of several cancers were examined in this chapter. While the endogenous chemical messengers and the corresponding organs were different yet specific, each type of cancer followed the same pattern outlined in the Hegedus Theory. The gene coding for the deactivating enzyme was altered to yield a non-functional protein. Consequently, the switch to stop cell division was lost.

It is a bitter irony that the very mechanism of defense against damage to the cell can become the instrument of the body's death.

CHAPTER 23

YOUR CURE

The Hegedus Theory involved no flash of genius, no magic, no incomprehensible leaps of faith. Plain old fashioned common sense was required to mold it. Finding the solution to the mystery of carcinogenesis was not all that difficult. Almost all of the pieces of the puzzle were available for assembly by any well educated scientist not blinded by conventional wisdom or fad theories fueled by the desire to satisfy the ego. Likewise, any such scientist not oppressed by the politics of cancer research could have seen the holes in the puzzle and done what I did to complete the picture. This explanation, the Hegedus Theory, was won at the cost of my livelihood as a cancer researcher - no small personal cost, but quite a bargain in view of the billions of dollars wasted over the past few decades chasing chimeras.

At one cancer convention I attended, a so-called colleague brow beat and then tried to ridicule me into revealing my theory on carcinogenesis. A friend of his, a Ph.D. immunologist, quietly observed and heard every word of this conversation. After it was over, this immunologist pulled me aside, but not to congradulate me for my accomplishment. He told me straight out, "I don't know if you really have unravelled the mechanism of cancer or not. But if you did, you'd better protect yourself." Then we started to talk. Why did I need protection for knowledge that should be available to everyone? I asked him casual questions about his current research projects. He told me he had gotten out of cancer research because he realized that the discipline of immunology, which constantly seeks to isolate new interferons, interleukins, and monoclonal antibodies, *could never cure cancer.*

In this book, when I described the cruelties inflicted on young scientists by the fascist regime in power now, it is from personal, painful experience. I lived it all, and I have tried to relate it to you. The pain, anguish, and suffering you may experience from this disease is shared by me. I lost my mother to cancer. As a call to moral duty, I have revealed the secrets of a perverse and violent subculture which, before this writing, was considered a pure, gentile "sacred cow" of our culture and the world alike: cancer researchers.

What happens next? I will undoubtedly be attacked for writing this book. I will be accused of grandstanding for bringing what is supposed to be an extremely technical and incredibly complicated scientific issue directly to the people, of trying to win popular support by appealing to emotion, of circumventing close scientific scrutiny of my work, and avoiding peer review. All of these predictable accusations are false. In fact, I challenge anyone to face me in public debate.

Admittedly, part of my motivation for writing this book was to spark an emotional response in you for an anger that is truly justified. But of what use is anger other than to drive someone to rectify its cause. You, the reader, may have long suspected that something is rotten about cancer research, but you couldn't quite place your finger on it. In shrouds of mystique related in incomprehensible jargon, cancer researchers have kept their scam in business. Now that you understand their secrets, you can rectify the cause.

I expect you to get very emotional about the needless waste of money and especially lives, animal and human, by the cancer research establishment. I have brought what are claimed to be highly complex and scientifically sophisticated concepts of cancer directly to you because you deserve to know. The basics of carcinogenesis are not difficult to understand. Any difficulty you have in grasping the concepts presented in the Hegedus Theory is a reflection of my ability as a teacher and writer rather than yours as a student.

My detractors will be partially correct about the lack of peer review concerning my work. But what was my recourse for this alleged acknowledgement of my work? On the one hand, my results were mostly suppressed when I made every effort to work within the cancer research establishment. On the other hand, I have been effectively ostracized from my hard earned profession, leaving me with either of two options. One, I take up another trade to earn money, in which case I leave you hostage to cancer and the cancer research establishment. Or two, I could fight back for my legitimate recognition as a scientist and for the rights of all scientists, and thereby, the public. I opted to fight back via this book!

Some will claim that the Hegedus Theory is unoriginal, others that it is unsupportable. To the first group I say, you are like the thief who calls all those around you crooks to mask your vile plan of theft. I welcome the comments from the second group. The Hegedus Theory is a theory in the finest sense, but currently a theory, nonetheless. It is firmly supported by scientific principles established by other scientists before me and well documented data. I did not work in a vacuum. This theory explains carcinogenesis better than any current or past alternative theories. It works. I (and you) need additional unbiased, imaginative research to further support and refine it in order to develop drugs that actually cure this disease.

But there is a catch, a Catch-22. The funding for research goes only to the cancer research establishment to support research leaders who have built careers on pursuing whimsical pipe dreams. They are threatened, not gladdened, by the reality that a major battle in the war on cancer has already been won.

The very people who would welcome the chance to pursue work (i.e., producing anti-cancer drugs) based on the Hegedus Theory will, like me, be deprived of the opportunity and funds to do so. Likewise, any victory in this battle, the full elucidation of the mechanism of cancers, is only a prelude to the complete victory in this war - cures for cancers. This work holds the best hope for cures, for specific drugs that will destroy each cancer as it appears in animals and humans.

At the risk of sounding hackneyed, I think the conduct of the war on cancer over the past few decades can be best compared to the difference in execution of the Vietnam War in contrast to the recent Persian Gulf War. In essence, the United States entered the Vietnam War motivated by principle and determination to stop Communism dead in its tracks. Under those circumstances, the cost in money and lives was necessarily paid. Over time, participation in this war slowly and subtley changed. It became a holding pattern. Tremendous amounts of money and horrendous loss of life was the consequence of the decision to perpetuate the war instead of winning it. There were many reasons for this decision, mostly political and at the time seemingly defensible.

Until the waning days of involvement, the Vietnam War enjoyed popular support - emotional, moral, and financial (via taxes). The majority of the citizenry gave its support because the basic principles (freedom and democracy) are noble and they *thought* that the war was being waged to be won. When enough people realized the commitment was to the status quo instead of victory, the expenditure of vast quantities of money and lives became indefensible. This finally precipitated the end of U.S. involvement. The parallels between this war and the war on cancer as it has been waged to this date are obvious.

Fortunately, the American people and their government have retained the capacity to learn from their mistakes. In 1991, the Gulf War was entered in defense of these same principles with the commitment to win, quickly and decisively. The government was well aware that popular support of a protracted conflict could not be sustained. The best resources and tactics available were brought to bear. Perhaps more importantly, unconventional and imaginative tactics were exploited as well, even to using snow plows on tanks to simply and effectively just fill in the reportedly impregnable desert trenches with sand. Every viable option was weighed and used to get the job done as quickly as possible. Why? Because loss of public support was counted equal with the costs in money and lives.

Now it is time for you to withdraw your emotional, moral, and financial support for the war on cancer as it has been ineffectively and inefficiently waged for the past 30 plus years. For the ruling regime, maintaining the status quo has become more important than making the commitment to win. I am here to win. Join me so that we all can win. The status quo keeps the powerful of the cancer research establishment fat and happy, at your expense. They have enjoyed high salaried careers and the admiration of the general public, while billions of dollars are wasted and millions of people suffer and die each year. These losers have little motivaion to win.

It is time for you, the people who provide the funds and suffer the deadly consequences, to withdraw your support and to insist on a commitment to winning. As long as the current research dictators remain in office, there is no incentive or hope for curing cancer. Only you have the power to force a change. First, contact your congresspersons and Senators. Tell them to read this book. Tell them you want an accounting of how your tax money has been spent to support cancer research. Put pressure on the legislators and regulators to allow us to take the politics out of cancer and the power from selected heads of cancer research programs. Second, simply stop giving your money to cancer research organizations. They are not the only game in town. They certainly are not the honest game they would have you to believe. Make sure they understand that you have had your fill of how the money they collect from you is being spent. The small amount of good they may accomplish does not justify the waste they support. The current corrupt system feeds on ego and money. The needless loss of lives obviously has not and will not change it. Only you, who provide the money and inadvertently reinforce their egos, can force this change. The power is within you. Act on it!

Someone has to lead the fight against cancer, but not the well documented failures currently controlling research. There have to be laws and procedures instituted to prevent the theft of ideas from talented subordinates, to prevent the wanton destruction of their careers by the powerful. Innovative or nontraditional ideas need to be fostered and rewarded, not suppressed or stolen. The authors of good ideas should be given the resources to pursue them, but the current system quickly breaks down in the hands of the ruling body now in control as an opportunity to steal new ideas that they are intellectually incapable of bringing to fruition. Innovative researchers should not be subjugated to the egos of the simple minds taking the form of supervisors or alleged colleagues who intellectually are inadequate to pursue new concepts. This all may sound impractical and naively idealistic. But with the stakes so high (your life), what other viable alternatives do we have? Perhaps more than anything, the past 50 years of cancer research has uncovered what *doesn't* work. Not only the misguided theories and treatments, but the entire infrastructure of the cancer research industry is a glaring testament to this *failure*.

Cancer as a disease can be cured. Once I and talented colleagues of mine are given the opportunity, it will be. And you will live long enough to see it. I have devoted more than 15 years of my life to defeating this menace called cancer. I am angered to report that the politics of research is a considerably more formidable obstacle than understanding and curing this disease. This book was written to give you the real picture of cancer, what is being done about it, and what can be done about it. If allowed, I will win this campaign against cancer. The only obstacle that remains is the honest opportunity.

I hope that this book has not depressed or dismayed you. It was designed to leave you with a heightened, profound understanding of this disease so that you would never again be forced to choose between false hope and no hope. If you look me straight in the eye and ask, "Can <u>you</u> cure cancers?" I'll look straight back into your eyes and answer, "<u>Yes</u>, I can."

You have been fed for years by the cancer research establishment with false hope - the promising new leads and the breakthroughs just around the corner. This will never be realized unless there is fundamental, extensive change. You can not lose hope. But of what use is hope without performance by the ones you have entrusted to cure cancer. Get mad. As mad as I am. We can and must unite forces to destroy this enemy. Obviously, cancer is one of them. However, it is equally important to rid all societies of another enemy, the dictators that decreed it is more important for you to suffer and die than it is for their egos to be bruised.

Get active. Stop believing that this disease is being cured or that any major new advances have been made against cancer. Do not give your hard earned dollars to support the egos of the "experts" in cancer. You have been conned. Now that you have read this book, don't let it ever happen to you again.

BIBLIOGRAPHY

MAGAZINES
Acta Cytological
Advances In Cancer Research
Advances In Experimental Medicine and Biology
American Journal of Pathology
American Journal of Physiology
Archives of Microbiology
Bulletin du Cancer
Cancer
Cancer and Metastatis Reviews
Cancer Letters
Cancer Research
Cancer Treatment Reports
Carcinogenesis
Cell, The
European Journal of Cancer, The
FEBS
Federation Proceedings
Infection and Immunology
Journal of Cellular Physiology
Journal of Clinical Pathology
Journal of Histochemistry and Cytochemistry
Journal of Immunology
Journal of Molecular Biology
Journal of the National Cancer Institute
Journal of Pathology
Lancet
Life Sciences
Lung
Molecular and Cellular Biochemistry
Molecular and Cellular Biology

Neoplasma
Proceedings of the National Academy of Science
Science
Scientific American
Toxicologic Pathology
Ultrastructural Pathology

SCIENTIFIC JOURNALS
Analytical Biochemistry
Analytical Chemistry
Archives of Biochemistry and Biophysics
Biochemica et Biophysica Acta
Biochemical and Biophysical Research Communications
Biochemical Pharmacology
Biochemistry
Biochemistry Journal
British Journal of Pharmacology
Canadian Journal of Biochemistry and Physiology
Chemical-Biological Interactions
Chemistry Reviews
Clinical Chemistry
Clinical Pharmacology and Therapeutics
Drug Metabolism Review
Enzymologia
International Journal of Biochemistry
Journal of the American Chemical Society
Journal of Biological Chemistry
Journal of Medicinal Chemistry
Journal of Organic Chemistry
Journal of Pharmaceuticals and Pharmacology
Journal of Pharmaceutical Sciences
Organic Chemistry
Proteins
Technometrics

Zenobiotica

TABLOIDS
The Journal of the American Medical Association
The New England Journal of Medicine

TEXTBOOKS
Basic Human Physiology: Normal Functions and Mechanisms of Disease, 2nd edition, Arthur C. Gurton, M.D., Ph.D. W.B. Saunders, Philadelphia, London, Toronto, 1977.

Biochemistry: A Problems Approach, William B. Wood, John H. Wilson, Robert M. Benbow, and Leroy E. Hood, The Benjamin/Cummings Publishing Company, Menlo Park, California, 1974.

Biology Today, 2nd edition, edited by David Kirk, CRM Random House, New York, 1972.

Cell and Molecular Biology, 7th edition, E.D.P. DeRobertis, M.D., and E.M.F. DeRobertis, M.D., Ph.D., Saunders College, Philadelphia, 1980.

Chemical Principles, 3rd edition, William L. Masterton, and Emil J. Slowinski, W.B. Saunders Company, Philadelphia, London, Toronto, 1973.

Clarke's Isolation and Identification of Drugs in Pharmaceuticals, Body Fluids, and Post-Mortem Material, 2nd edition, edited by A.C. Moffet et. al., The Pharmaceutical Press, 1986.

Clinical Toxicology of Commercial Products: Acute Poisonings, 4th edition, Robert E. Gosselin, M.D.,

Ph.D., Harold C. Hodge, Ph.D., D.Sc., Roger P. Smith, Ph.D., Marion N. Gleason, M.Sc., William & Wilkins, Baltimore/London, 1976.

Cutting's Handbook of Pharmacology: The Actions and Uses of Drugs, 6th edition, T.Z. Csaky, M.D., Appelton-Century-Crofts, New York, 1979.

Drug Metabolism and Drug Toxicity, edited by Jerry R. Mitchell, M.D., Ph.D., and Marjorie G. Horning, Ph.D., Raven Press, New York, 1984.

Drug Metabolism: Chemical Aspects, Bernard Testa and Peter Jenner, Marcell and Dekker, Inc., New York and Basel, 1976.

Enzyme Kinetics: Behavior and Analysis of Rapid Equilibrium and Steady-State Enzyme Systems, Irwin H. Segel, John Wiley & Sons, New York, Chichester, Brisbane, Toronto, 1975.

Foundations of Behavioral Research, 2nd edition, Fred N. Kerlinger, Holt, Rinehart and Winston, Inc., New York, Chicago, 1973.

Fundamentals of Physical Chemistry, Samuel H. Maron, and Jerome B. Lando, Macmillan Publishing, Co., Inc., New York, 1974.

Goodman and Gilman's: The Pharmacological Basis of Therapeutics, 7th edition, edited by Alfred Goodman Gilman, M.D., Ph.D., Louis S. Goodman, M.A., M.D., D.Sc. (Hon.), Theodore W. Rall, Ph.D. D.Med. (Hon.), Ferid Murad, M.D., Ph.D., Macmillan Publishing Co., Inc., New York, 1985.

Guide to General Toxicology, A, edited by Freddy Homberger, John A. Hayes, Edward W. Peliken, Karger, 1983.

Handbook of Basic Pharmacokinetics, 2nd edition, W.A. Ritscell, Drug Intelligience Publications, Inc., Hamilton, IL, 1980.

Instrumental Methods of Analysis, 5th edition, Hobart H. Willard, Lynne L. Merritt, Jr., John A. Dean, D. Van Nostrand Company, New York, 1974.

Introductory Statistics for the Behavioral Sciences, Douglas W. Schoeninger and Chester A. Insko, Allyn and Bacon, Inc., Boston, 1971.

Medicinal Chemistry: A Biochemical Approach, Thomas Nogrady, Oxford University Press, New York, Oxford, 1985.

Merck Index, The: An Encyclopedia of Chemicals, Drugs, and Biologicals, 10th edition, edited by Martha Windholz, Merck & Co., Inc., Rahway, N.J., USA, 1983.

Modern Concepts in Biochemistry, 2nd edition, Robert C. Bohinski, Allyn and Bacon, Inc., Boston, London, Sydney, Toronto, 1976.

Modern Experimental Organic Chemistry, 4th edition, Royston M. Roberts, John C. Gilbert, Lynn B. Rodenwald, Alan S. Wingrove, Saunders College Publishing, Philadelphia, New York, 1985.

Organic Chemistry, 3rd edition, Robert Thorton Morrison and Robert Neilson Boyd, Allyn and Bacon, Inc., Boston, 1973.

Organic Synthesis: The Disconnection Approach, Stuart Warren, John Wiley & Sons, Chichester, New York, Brisbane, Toronto, Singapore, 1982.

Statistics In Medicine, Theodore Colton, Sc.D., Little, Brown, and Company, Boston, 1974.

Statistical Package for the Social Sciences, Norman H. Nie, et. al., McGraw-Hill Book Company, 1975.

Sterile Dosage Forms: Their Preparation and Clinical Aplications, 2nd edition, Salvatore Turco, M.S., Pharm. D., and Robert E. King, Ph.D., Lea & Febiger, Philadelphia, 1979.

Taber's Cyclopedic Medical Dictionary, 12th edition, edited by Clayton L. Thomas, M.D., M.P.H., F.A. Davis, Philadelphia, 1973.

Teratology: Principles and Techniques, edited by James G. Wilson and Josef Warkany, The University of Chicago Press, Chicago and London, 1972.

Wilson and Giswold's Textbook of Organic Medicinal and Pharmaceutical Chemistry, 8th edition, edited by Robert F Doerge, Ph.D., J. B. Lippincott Company, Philadelphia, Toronto, 1982.

GLOSSARY

A

Activated carcinogens - Metabolites formed from enzymes that have altered the structure of a chemical such that it reacts with biomolecules, including DNA.

Activated metabolites - Reactive compounds resulting from the enzymatic transformation of a relatively non-reactive chemical.

Active site directed agents - Chemicals that molecular biologists claim target a specific gene at a specific site.

Adenocarcinoma - A malignant, or rapidly growing, type of cancer arising from outside surface cells of an organ.

Adrenalin - A neurotransmitter produced primarily in the small glands above the kidney, called the adrenals, which stimulate the heart and the respiratory rates.

Adrenal glands - Organs where adrenalin is produced.

Albumin - A protein in the blood that carries chemicals, including chemicals that are carcinogens.

Alkylating agents - Highly reactive chemicals that permanently bond to all biomolecules without enzymatic activation.

Alkylation - A reaction that bonds a portion of one molecule to another another chemical, usually DNA,

RNA, and proteins.

Alpha$_1$-Acid glycoproteins - Complex biomolecules with a number of sugar molecules attached to it that may help the immune system to form antibodies.

Alveolar type II cells - Air cells deep in the lung where oxygen and carbon dioxide are exchanged.

Amino acids - Organic acids that contain the amino (-NH$_2$) functional group. There are about 21 amino acids that are strung together in various combinations to form proteins, some of which are enzymes and pharmacological receptors.

Amino acid functional group - The reactive -NH$_2$ moiety found in biomolecules and organic chemicals.

Analgesics - Medicines that relieves pain.

Analogue - Any one chemical in a series of compounds with similar structures.

Analytical chemists - Scientists who extract, identify, and quantitate chemicals found in air, biological samples, soil, and water.

Anaphylaxis - A state of severe clinical shock which is potentially fatal as a result of exposure to a chemical. An extreme allergic reaction to a bee sting is an example of anaphylaxis.

Anemia - A condition in which there is a reduction in the number of red blood cells or in the volume of blood, or both.

Anorexia - A disorder that results in a severe loss of appetite.

Antibiotic - Natural and syntheitic chemicals that inhibit the growth of microorganisms.

Antibody - A protein developed by the immune system as a response to invasion by a microorganism, or other large bodies such as proteins and DNA fragments.

Antigen - A foreign substance, like microorganisms, that induce the immune system to produce antibodies.

Antihistamine - A medicine that counters the effects of histamine. It is found in a number of over the counter cold remedies and allergy medications.

Antimetabolite - A chemical that can be useless or harmful which has a structure similar to and substitutes for a chemical necessary for life of the organism.

Antimitogen - A chemical which inhibits or prevents the division of cells.

Antipsychotic - A class of drugs used to treat severe mental illness.

Antiviral - An agent that destroys viruses.

Aromatic - A portion of a molecule that has chemical characteristics similar to benzene.

Atrophy - A wasting away of an organ or tissue due to lack of nutrients or disuse.

Autoclave - An apparatus used for sterilization by steam pressure and a temperature above 250° F.

Autocoid - Substances produced by the body to help regulate its proper chemical functions. Examples of autocoids are serotonin and histamine.

B

Bacteria - A plantlike microorganism.

B-cell - One of the white blood cell types that makes antibodies against invading microorganisms which are formed in the liver and spleen several months after birth.

Biochemistry - The study of the chemistry of living things.

Biochemists - Scientists who study the chemical reactions pertaining to the necessary processes of life.

Biomolecules - Substances that comprise the living system, eg., enzymes and genes.

Biopsy - Excission of a small piece of tissue for light microscopic identification.

Bone marrow depression - Chemical inhibition of soft tissue in the long bones responsible for the production of red blood cells.

Bone resorption - Shrinkage of the bone.

C

Cancer - A rapid, uncontrolled growth of a cell type in an organ.

Cancer-suspect agent - A chemical that is administered to an animal along with a known carcinogen to induce cancer. A favorite tactic of witch hunters who call themselves cancer researchers to link a chemical or activity to cancer.

Carcinogen - An agent, chemical or radiation, that initiates the formation of cancer.

Carcinogenesis - The process of biochemical events leading to the origin of cancer.

Cardiac - Pertaining to the heart.

Cardiomyopathy - A disease or abnormal condition of the heart.

Cardiotoxicity - Poisonous effects on the heart caused by chemicals.

Cartel - An association designed to establish an international monopoly to control the access and dispensation of knowledge.

Catalytic pocket - The site in an enzyme that alters chemicals.

Cell - The most basic unit of complete life.

Cell biology - The study of processes in the most basic

unit of life called the cell.

Cell wall (membrane) - The outermost part of the cell that retains its contents and excludes most chemicals.

Chemical - A substance which has one or more elements bound to each other.

Chemical carcinogens - Substances that cause cancer.

Chemical messenger - A compound that the body produces in one organ which is released into the bloodstream to stimulate activity in other organs.

Chemistry - The science that seeks to understand the interaction of atoms and molecules with each other.

Chemotherapeutic agents - Drugs which are intended to kill the invading microorganism or cancer cells, while leaving unharmed the other cells in the body.

Chemotherapy - The process of using chemicals in an attempt to kill specific cell types and leaving the others unharmed.

Clara cells - Type of lung cell that is responsible for the metabolism of foreign chemicals.

Colorimetric method - A technique used by chemists to identify and quantitate the amount of a chemical present in the sample. The chemical analyzed for reacts with another chemical to produce a specific color. The intensity of the color corresponds to the amount, or concentration, of chemical in the sample.

Concentration - The amount of chemical dissolved per unit volume of liquid.

Congeners - Chemicals in a series that have only slight alterations in the structure of each.

Coomassie Brilliant Blue G - A dye used to detect and quantitate proteins.

Cytoplasm - The substances inside the cell other than the nucleus.

Cytotoxicity - A substance that is poisonous to the cell.

D

Dermatitis - An inflamation of the skin experienced by itching, redness, and various cuts in the skin.

Differentiation - The process whereby the cell makes specialized structures responsible for the special function of that cell type.

Digest - In reference to gene extraction, it is the fragments of DNA left after endonuclease enzymes snip the molecule into pieces.

Dosage regimens - The amount of several drugs given to a patient according to a strict time schedule.

Dose-Response curve (relationship) - The measured effects of a chemical on an organism based on the concentration (amount) of chemical given.

Dysplasia - An abnormal development of a tissue.

E

Electron microscope - An instrument used to view samples at very high magnifications, up to 200,000 times.

Endogenous chemical messengers - Substances produced by the body in one organ that are secreted into the bloodstream to stimulate activity in another cell type or organ.

Endometrial cancer - Tumors of the lining in the uterus.

Endoplasmic reticulum - A canal-like structure in the cell which contains metabolic enzymes, the P-450's for example.

Enzymes - Complex proteins produced by the body that alter the structure of chemicals. Enzymes are relatively specific for a given chemical.

Enzyme inhibitors - Compounds that block the enzyme's ability to alter chemicals.

Epithelial tumors - Cancers that arise from cells that line the cavities, tubes and passages which lead to the outside of the body. The inside of the nose is an example of where epithelial cells are located.

Epstein-Barr virus - A virus that allegedly causes leukemia in cats.

Estrogens - Several steroid compounds called hormones

produced in the ovary that impart female charateristics like breast development and the menstrual cycle. Estradiol and estrone are examples of estrogens.

F

Fad theories - Notions that have arisen to explain cancer based upon reactions to current political and social ideologies with no basis in fact or scientific principles.

Feedback inhibition - The process whereby an enzyme's activity is reduced, or even stopped, by the end product of its own catalytic reaction.

Fluorescent dye technique - A method to view the attachment by monoclonal antibodies to tissue samples using a light microscope. When light hits the sample, yellow-green dots appear at the sites of the monoclonal antibody-dye attachment.

G

Gamma ray - A high energy electromagnetic entity, similar to the x-ray, emitted by some radioactive compounds. They have greater penetrating power than alpha and beta particles.

Gel electrophoresis - The movement of charged particles, eg. DNA fragments and proteins, through a gelatin-like substance which has an electrical charge passed through it of typically 500-1,500 volts, but a low current of about 100 microamps.

Genes - A DNA fragment that codes for a protein which could be an enzyme or pharmacological receptor. They contain hereditary information.

Gene therapy - A technique that tries to correct a flawed sequence of nucleic acids in the DNA fragment. In cancer research, this means that the alleged oncogene is turned off or replaced by the correct gene.

General protoplasmic poison - A chemical that indiscriminately kills all life forms which come into contact with it. Most anti-cancer drugs act by this mechanism.

Genetic code - The sequence of nucleic acids that comprise the genes.

Genetic engineering - A process whereby a DNA fragment, either synthetic or from one organism is placed into the genetic code of another organism which changes the traits of the latter.

Genitourinary cancer - Tumors found in the reproductive organs, including the uterus.

Glioma - A cancer of neurons.

Guru - A spiritual leader or guide.

H

Hairy cell leukemia - A type of white blood cell cancer in which the membrane has many hair-like protrutions covering it.

Hepatic toxicity - Damage to the liver caused by chemicals.

Heredity - Traits and characteristics developed by

individuals due to their various genes passed to them from ancestors.

High performance liquid chromatography (HPLC) - A technique to separate chemicals by passing them through a specially packed column at pressures between 500 and 2,000 pounds per square inch (PSI).

Hormone - A substance synthesized in a gland or organ that is secreted into the bloodstream to stimulate activity in other organs.

Hormone antagonists - Drugs that block the effects of a hormone at its pharmacological receptor.

Hydrophilicity - The characteristic of a molecule that makes it water soluble.

5-Hydroxytryptamine - Another name for the chemical messenger, serotonin, synthesized and secreted by neuroendocrine cells and neurons in the brain.

5-Hydroxytryptophan - The amino acid precursor used to synthesize serotonin.

Hyperplasia - An excessive proliferation of normal cells in an organ which often is misdiagnosed as a cancerous growth. This abnormal growth is not harmful and is usually reversible in that it will recede with time.

I

Immune system - That part of the body comprised of white blood cells, lymph glands, and the thymus which is responsible for attacking and destroying microorganisms that invade the body.

Immunologist - A researcher who studies the immune system.

Immunoregulators - Agents that determine the number and types of white blood cells produced in response to infections.

Inhibitor - A chemical that hinders or stops the activity of an enzyme.

Intercalation - An extraneous chemical that gets inserted between two other chemicals.

Interferons - Proteins produced by white blood cells in response to viral infection.

Interleukins - Proteins formed by the white blood cells in response to infection by microorganisms.

K

Kaposi sarcoma - A type of skin cancer arising from the underlying skin layers.

L

Laser - A high energy light beam that is very concentrated.

Leukemia - A rapid and abnormal proliferation of white blood cell, leukocytes, coming from the blood-forming organs, bone marrow, lymph glands, and spleen.

Leukopenia - Abnormal decrease in the number of white

blood cells.

LD$_{50}$ - Lethal dose of a chemical that kills almost immediately 50% of the animals tested.

Light microscope - An instrument that magnifies samples which are illuminated by an ordinary light bulb. Upper limit of magnification is about 1,500 times.

Lipophilicity - A characterisitic of a molecule that makes it fat soluble.

Liver toxicity - A term used interchangably with hepatic toxicity that refers to substances which cause damage to the liver.

Long-term study - A set of experiments that typically last one to five years.

Lymphoma - A cancer of the lymphatic system. Hodgkin's disease is an example.

Lysosomes - A group of powerful digestive enzymes contained in intracellular sacs that are released when the cell is infected with a microorganism. They comprise part of the cell's defense system which destroys proteins and carbohydrates. Usually, they continue to act until they destroy themselves.

M

"Magic bullet" - This is a term that refers to drugs which target a specific cell type much like the group of drugs called sulfonamides effective against syphilis developed by Dr. Ehrlich, an organic chemist.

Malignant (malignancy) - A rapidly growing form of a tumor with the potential to produce death.

Mass spectrophotometer (MS) - An instrument used by chemists to identify the structure of a chemical by exposing it to a high energy electrical charge which fragments the molecule into a characteristic or "fingerprint" pattern.

Mechanism of action - An explanation, based on scientific principles, of how a chemical or other agent works.

Medicinal chemist - A scientist who designs and develops new drugs, or medicinals.

Melanoma - A pigmented mole on the skin that is malignant.

Metabolism - Structural alterations of chemicals due to the action of enzymes.

Metabolite - The newly formed product of the enzyme's action.

Metallothioneins - Complex proteins that have metal ions bound to them.

Metastasis - The movement of cells from one body part to another in the traditional understanding of cancer. More accurately, they are tumors that form in parts of the body other than the original, or primary, organ.

Metastatic tumor - A cancer that forms away from the organ in which it was originally detected.

Methylene chloride - A common chloro solvent used by chemists to extract chemicals from biological and aqueous (water) samples, that has the structure CCl_2H_2.

Methylation - A reaction that bonds a CH_3 (methyl) moiety to another chemical. In cancer research, this chemical is usually DNA.

Methyl isocyanate - A highly reactive and toxic chemical that puts a methyl moiety onto other biomolecules. It has the structure CH_3-N=O.

Microsomes - A group of enzymes responsible for the metabolism of chemicals located in an organelle which looks like a network of canals when viewed by the electron microscope.

Microtubules - Microscopic tubes of protein that form during cell division to help push apart the two new cells.

Mitochondria - Organelles inside the cell that appear cigar shaped when viewed by the electron microscope where most of the chemical energy is produced.

Mixed function oxygenases (P-450's) - Enzymes, found in the liver and other organs, which transform chemicals by incorporating oxygen atoms into the original structure. They sometimes are referred to as hemeproteins and there are several forms.

Molecular biology - The discipline which seeks to extract and locate genes with the intent to change the genetic code of an organism.

Molecules - The smallest unit a chemical can be divided into while retaining all of its special characteristics.

Monoamine oxidase (MAO) - One of two, A and B, forms of an enzyme that de-activates several endogenous chemical messengers and foreign chemicals by removing an amino ($-NH_2$) moiety to form the corresponding aldehyde.

Mutation (mutagenesis) - A permanent change in the genetic code caused by either chemicals or radiation.

Myelogenous leukemia - A cancer of white blood cells that originate in the bone marrow.

Myeloma - A tumor originating in the blood producing portion of the bone marrow.

Myelopathy - A deterioration of the spinal chord.

N

Natural Killer (NK) cells - White blood cells that seek out and destroy invading microorganisms.

Natural products - Chemicals extracted from animals, plants, and microorganisms that are examined for utility as drugs.

Necrosis - Localized death of a tissue.

Neuroblastoma - A malignant form of a cancer derived from nerve cells throughout the body that stimulate some organs like the heart and lungs seen mainly in children and infants.

Neuroendocrine cells - Those cells in the body which produce, secrete, and deactivate endogenous chemical messengers, like serotonin, that stimulate the activity of neurons and some organs.

Neurons - Another name for a nerve cell.

Neurotoxicity - Poisonous effects to nerve cells caused by chemicals.

Neurotransmitter - Chemicals synthesized by nerve cells to facilitate communication between themselves.

Nitrogen mustards - Chemical variants of the original "nerve" gasses used during World War I. They are highly reactive general protoplasmic poisons.

Nitrosamines - A group of chemicals that contain the N-N=O moiety, many are carcinogens.

Nuclear magnetic resonance (NMR) spectrophotometer - An instrument used by chemists to identify the structure of a chemical by taking advantage of the reverse spin of about one in a billion hydrogen atom nuclei in the sample that are detected and amplified. Currently, this technique is called magnetic resonance imaging (MRI).

Nucleic acid - A group of five chemicals that are strung together in various sequences to produce genes and RNA.

Nucleus - The central part of the cell which contains all of the genes and is surrounded by a special membrane.

"Nude" mouse - A strain of mouse which has been genetically engineered to be devoid of its immune system.

O

Oncogenes - A fabrication by molecular biologists which are alleged genes that code for any one of a variety of cancers. There is no scientific evidence to support this.

Organ - Part of the body that provides a special function.

Organelle - A specialized structure inside the cell that performs a specific function. Examples of organelles are mitochondria and microsomes.

Osteoblasts - A type of cell which aids in the formation of bones.

Osteoporosis - An increase in the spaces between bone cells, or a softening of the bone.

Oxygen electrode - An electronic devise that measures the concentration, i.e., amount, of oxygen in water.

P

Papilloma - Another name for a wart.

Parethesis - An abnormal sensation such as numbness, prickling, or tingling without any objective cause experienced by nerves in the body.

Pathologist - A researcher specialized in the diagnosis of a disease in tissues removed during operations or post-death samples.

Peripheral nervous system - The entire nervous system located throughout the body other than the brain and the spinal chord.

Pharmacognosy - The science of extracting chemicals from animals and plants that may have therapeutic value.

Pheochromocytoma - A type of tumor that originates in the adrenal glands that produces excess amounts of adrenalin.

Phlebitis - Inflamation of the viens.

Phosphagenes - A group of highly reactive compounds similar to nitrogen mustards that contain the element phosphorus.

Phosphodiesterase - An enzyme that converts a nucleic acid attached to a phosphate group to a cyclic arrangement of the chemical. In this cyclic arrangement, the chemical is an energy releasing compound.

Phosphorus - A non-metallic element essential to all life forms, but toxic in the excess.

Piperonyl Butoxide (PBO) - The chemical most often used by biochemists to inhibit the activity of the mixed function oxygenases (P-450's).

Polycyclic Aromatic Hydrocarbons (PAH) - A group of compounds found in the exhaust fumes of vehicles which have several fused ring systems and a distinctive noxious odor. Most of these chemicals are potent carcinogens.

Polymerase - An enzyme that synthesizes new DNA sequences from template, or original, fragments.

Progesterone - One of several hormones produced by the body (endogenous) that stimulates a variety of functions in the female reproductive system.

Prostatic acid phosphatase - An enzyme of the prostate gland.

Proteins - Any and all sequences of amino acids strung together in every conceivable order. Proteins can be enzymes, pharmacological receptors, structrual supports for cells, or nonsense fragments.

Pulmonary - That which pertains to the lung.

Pulmonary embolism - A blood clot that finds its way to the lung. It is a potentially lethal condition.

R

Radiation - Intense emission of matter or rays from a source that has the potential to cause chemical alterations in the biomolecule it hits.

Radioactivity - Emission of radiation which can be alpha and beta particles, or gamma rays.

Receptor - A protein which can be on the membrane of the cell or at several locations within the cell that has an affinity for the temporary binding of specific chemical structures.

Renal sarcoma - Tumors in the kidney.

Respiratory tract - The apparatus in our bodies used to breathe which includes the nose, tonsils, larynx, pharynx, trachea, and bronchi of the lung.

Restriction endonucleases - Enzymes that recognize very specific sites of short DNA and RNA fragments to cleave the molecules at these locations.

Restriction enzymes - Another name for restriction endonucleases.

Retinoblastoma - A malignant form of eye cancer found most often in infants.

S

Sacrifice - A word medical researchers use to describe killing and butchering animals.

Scientificese - The jargon of medical researchers which sounds like the common language of the country, but, by design, is incomprehensible to the average person.

Sea cucumber - A long, slow moving, bottom dwelling animal found in tropical oceans. It gets its name from resembling the shape of a common vegetable, the cucumber.

Selective toxicity - An agent that kills only a specific cell type without harming the rest of the organism.

Septicemia - Another term for blood poisoning by an infection of microorganisms.

Short-term study - Experiments with animals, including humans, that last between a few hours and a few days.

Side effect - An undesirable or unexpected reaction by an organism to a drug, eg., allergic reaction to aspirin.

Sinigrin - A chemical found in mustard seeds, horseradish, and certain vegetables like broccoli and cabbage which was decreed by cancer researchers to prevent the formation of cancer. There is scientific evidence to refute this proclamation.

Sociological studies - A set of experiments with people that try to determine the causal relationships between a series of events. A favorite battle cry of physicians is " Studies have shown".

Spectrophotometer - An instrument used to measure the amount (concentration) of a chemical in a solution by passing a light beam dialed to a very precise wavelength through the sample solution.

Statistics - A method to gather numerical data which is sorted, then analyzed and interpreted by using a series of mathematical equations.

Strain - Another term for breed of animal.

Structure-activity relationships - Understanding the effects of a series of drug analogues by examining the components of the molecule.

Suspect carcinogen - A chemical which does not cause cancer, but has been linked to the disease by witch hunters who want to make a quick name for themselves. These chemicals are given to laboratory animals along with a known chemical carcinogen to induce cancer as the typical modus operendus of researchers.

T

Tamoxifen - The proprietary name for a drug which blocks the entrance of estrogens into organs like the breast. Another name for this type of drug is hormonal antagonist.

Target organs - The tissue in the body toward which the greatest effect is exerted by a chemical, including a drug.

Taxol - The proprietary name for a group of chemicals extracted from the bark of the yew tree which is currently being touted as the "cure" for breast cancer. Fatalities in farm animals due to Taxol are common. Human fatalities result from gastrointestinal irritation, cardiac and respiratory failure. In short, it is another general protoplasmic poison heralded as a cure for cancer.

Thrombocytopenia - Abnormal decrease in the number of blood platelets. It is a potentially fatal condition because a patient can bleed to death from an ordinary cut.

Tissue - A group or collection of similar cell types and their intercellular substances which act as a unified whole to perform a specific function.

Tissue culture - A technique that keeps individual cells alive with a special nutrient broth housed in a carefully controlled environment.

Tobacco carcinogens - Chemicals found in all tobacco products that cause cancer in all animal species tested.

Tumor - A mass of tissue resulting from an abnormal growth.

Tumoricidal - Killing only cancer cells.

Tumor necrosis factor - A protein isolated from white blood cells which has been decreed by cancer researchers to allegedly kill only cancer cells. In tissue culture experiments and in clinical trials, it has consistently proven to be worthless against any type of cancer.

U

Ultrastructure - The fine network of organelles and structures inside the cell.

Ultraviolet (UV) light - Rays of light that eminate from beyond the violet rays of the visible specrum.

V

Vascularization - The process of manufacturing new blood vessels in tissues.

Virus - A microorganism which can not be viewed by a

light microscope. They are parasites which depend on nutrients found in the host cell for their metabolic and reproductive needs.

W

White blood cells - Cells in the blood stream that appear white when viewed by the light microscope. They patrol the body seeking and destroying invading microorganisms.